# UNFINISHED SYNTHESIS

# UNFINISHED SYNTHESIS

## Biological Hierarchies and Modern Evolutionary Thought

NILES ELDREDGE

New York   Oxford
OXFORD UNIVERSITY PRESS
1985

Oxford University Press

Oxford   New York   Toronto
Delhi   Bombay   Calcutta   Madras   Karachi
Petaling Jaya   Singapore   Hong Kong   Tokyo
Nairobi   Dar es Salaam   Cape Town
Melbourne   Auckland

and associated companies in
Beirut   Berlin   Ibadan   Nicosia

First Published in 1985 by Oxford University Press, Inc.,
200 Madison Avenue, New York, New York 10016

First issued as an Oxford University Press, Inc., paperback, 1988

Oxford is a registered trademark of Oxford University Press.

Library of Congress Cataloging in Publication Data
Eldredge, Niles.
Unfinished synthesis.
Bibliography: p. Includes index.
1. Evolution.   I. Title.
QH366.2.E53   1985   575   85-5008
ISBN 0-19-503633-6
ISBN 0-19-505574-8 (pbk.)

10 9 8 7 6 5 4 3 2 1

# Preface

This book is concerned with how we think about evolution—how we have been *trained* to think about evolution and how, in my opinion, we *should* be thinking about it. It is thus a book about the structure of evolutionary theory, at least as much as it is a book *of* evolutionary theory. I believe that the range of phenomena, problems, and theory at our disposal today remains pretty much as it was in Charles Darwin's day, with only the nota-ble addition of a formal and reasonable theory of inheritance now availa-ble. I start here with the works of Theodosius Dobzhansky (1937a and 1941, the first two editions of *Genetics and the Origin of Species*), Ernst Mayr (1942, *Systematics and the Origin of Species*), and George Gaylord Simpson (1944, *Tempo and Mode in Evolution*) to determine what actually is on that list of phenomena, entities, and processes that bear on an under-standing of evolution—and as support for my conviction that little new, in terms of the very basics of the structure of evolutionary thinking, has occurred since these books were written.

I further believe that the current version of the "modern synthesis" remains so unmoved by the data of systematics, paleontology, and large-scale ecology that we truly still face the very situation diagnosed by his-torian F. J. Teggart in 1925 just as the synthesis was beginning to emerge through the labors of R. A. Fisher, J. B. S. Haldane, and S. Wright: we still have a theory of evolution that is not directly addressed to the actual events of the history of life. It is not a major goal of this book to explain *why* this state of affairs is so; I shall be content merely to establish that it *is* so.

I ask but one thing of my readers: a willingness to think about evolution "as if" (as Lewontin 1980, p. 65, put it) certain positions on the nature of biological entities are true. The crucial issues in this book are for the most part ontological; that is, I explore the implications for evolutionary theory of the *possibility* that such items as species, monophyletic taxa, and com-munities may be construed as "individuals" in the sense of Ghiselin

(1974a). Indeed, I try mightily to persuade my readers that this is so and that, *if* so, the consequences for the structure of evolutionary theory are necessarily profound. There is little purpose to be served in reading this book without a willingness at least to entertain the possibility that the roster of biological entities—individuals in a formal, specifiable sense—that take part in the evolutionary history of life may not be complete in traditional evolutionary theory.

I write this book in the spirit of conviction that there must indeed be a single evolutionary process and that somehow all biological phenomena must be relevant to understanding that process. This is, of course, the very same spirit that motivated the synthesis in general and the works of Dobzhansky, Mayr, and Simpson that figure so prominently in my early pages here. It is they who have inspired the present inquiry.

*New York*                                                                                    N.E.
*January 1985*

# Acknowledgments

I have been pursuing this business of evolution for a number of years now. Most of my work has been done with others, either directly or through conversation or the more remote printed page. I want to mention a few of my colleagues here. My interest in trying to forge a more coherent liaison between evolutionary theory and the data of paleontology—a task that still seems so embryonic—is perhaps best known through the notion of "punctuated equilibria." The kernels of my hierarchical thinking lie in a portion of the paper that gave punctuated equilibria its name. That paper, of course, was written with Stephen Jay Gould. The ideas expressed here started there in large measure. But even more important, perhaps, than the mutual development of a system of thought was the example that Steve set for a number of us, even in the early days: fearlessly confronting the world of professional biology and, indeed, making it his oyster.

Closer to the present, many of the ideas I express here, particularly those on various aspects of hierarchy theory and its relation to evolution, come from close association with two biologists: Stanley N. Salthe and Elisabeth S. Vrba. I have written papers with each of them, developing various corners of what I take to be an exciting way of approaching evolutionary theory. Their specific contributions are more fully acknowledged within the text. Here I merely wish to thank them for working with me on the problem in general—and to ask them to forgive any departures from the still nascent orthodoxy that they may single out.

There are others, of course. Teachers and colleagues such as Norman Newell and Bobb Schaeffer, coauthors and colleagues such as Joel Cracraft, Michael Novacek, and Ian Tattersall, have all helped immeasurably. Some have even gone so far as to read earlier versions. I am grateful, too, to David Hull and John Damuth for their helpful comments on the manuscript. It has also helped for me to be privileged to work at the American Museum of Natural History, the very institution that contributed so much to the development of the synthesis.

The author is grateful for permission to reprint material from the following sources:

K. S. W. Campbell, *Oklahoma Geological Survey Bulletin 113*. By permission.

Theodosius Dobzhansky, *Genetics and the Origin of Species*. Copyright © 1937, 1941, 1951, 1982, Columbia University Press. By permission.

Niles Eldredge, *Introduction to Mayr 1942 (1982)*. Copyright © 1982, Columbia University Press. By permission.

James A. MacMahon et al., *Bioscience*, vol. 28, p. 701. Copyright © 1978 by the American Institute of Biological Sciences. By permission.

Ernst Mayr, *Systematics and the Origin of Species*. Copyright © 1942, 1982, Columbia University Press. By permission.

George Gaylord Simpson, *Tempo and Mode in Evolution*. Copyright © 1944, 1984, Columbia University Press. By permission.

George Gaylord Simpson, *The Major Features of Evolution*. Copyright © 1953, Columbia University Press. By permission.

# Contents

# UNFINISHED SYNTHESIS

# 1

# Approaching Complexity: Thinking About Evolution

"The basic problems of evolution are so broad that they cannot hopefully be attacked from the point of view of a single scientific discipline." (Simpson 1944, p. xv)

Evolution is a complex affair. The very diversity of definitions of biological evolution illustrates the variety of ways in which we think about the subject. Perhaps still the best general description of evolution is Darwin's "descent with modification," simply because it simultaneously brings to mind a pattern—a result—while hinting at an understanding of process, an underlying causative mechanism. Other definitions are less broad and thus fare less well as general descriptors of this thing we call evolution. I have characterized evolution as the proposition that all organisms are descended from a single common ancestor (e.g., Eldredge 1982c, based on a suggestion from N. I. Platnick in a personal communication). This is a serviceable (and testable) construct but one that emphasizes a systematist's concern for pattern. The utterly different and far more popular notion that evolution is "change in gene frequencies within a population" similarly emphasizes process through focusing on a completely different sort of pattern. Yet neither definition is "wrong."

The "modern synthesis" is a body of thought that grapples with the complexities of evolution. As does any good theory, the synthesis attempts to characterize the overall phenomenon and explain it in the simplest terms that seem appropriate and effective. The bewildering array of evolutionary process theories that had accumulated by the 1920s, where each biological discipline seemed bent upon establishing the primacy of its own phenomena and its own insights into processes, amounted to a net disarray for evolutionary biology. As Simpson wrote in 1944 (p. xv):

Not long ago paleontologists felt that a geneticist was a person who shut himself in a room, pulled down the shades, watched small flies disporting themselves in milk bottles, and thought that he was studying nature. A pursuit so removed from the realities of life, they said, had no significance for the true biologist. On the other hand, the geneticists said that paleontology had no further contributions to make to biology, that its only point had been the completed demonstration of the truth of evolution, and that it was a subject too purely descriptive to merit the name "science." The paleontologist, they believed, is like a man who undertakes

3

to study the principles of the internal combustion engine by standing on a street corner and watching the motor cars whizz by.

The situation now, though far from ideal, is vastly improved, and for that we may thank the synthesis.

Although, as we shall shortly see, the early statements of the synthesis were indeed rather more pluralistic than later formulations in their choice of both patterns and explanatory processes for inclusion within the theory, from the start (say, arguably, Dobzhansky 1937a) the architects of the synthesis saw the origin, maintenance and further modification of adaptations through natural selection as the central, if not sole, evolutionary process. Thus, the synthesis approached a bewildering array of both theory and the genuinely diverse melange of natural phenomena we generally include under the term *evolution* in a spirit of simplification; perhaps the biological world in general, and the evolutionary process in particular, was not as complex as it had seemed to be.

But the price of this simplification has struck some of us as a bit steep. Geneticist Hampton Carson (1981, p. 773), in a letter written to *Science* in response to Lewin's (1980) report on a conference on macroevolution, stated the conventional view of the synthesis:

The fossil record says eloquently that profuse evolution has indeed occurred over millions of years, but the data just aren't sensitive enough to analyze evolutionary kinetics. This is the province of the evolutionary geneticist who works with descent and change in populations of present-day organisms.

In contrast, the central theme of this book asserts that there are additional entities and processes—hence "kinetics"—germane to any complete conceptualization of evolution and beyond the conventional purview of evolutionary geneticists.

Of the many things being said about evolution in the 1980s, the notion that all positions remain subsumed under the general banner of the modern synthesis (as claimed, for example, in Stebbins and Ayala 1981) is surely one of the more provocative. It raises the simple question: Just what *is* the synthesis, anyway? I pursue this question at length through careful analysis of what I take to be the most significant statements of the synthesis. It has long been clear that genes, species, and phyla all fit somewhere into the evolutionary scheme of things; early writings of the synthesis made just this point and attempted to reintegrate what had become a confusing welter of mutually inconsistent theories marshaled by various camps of geneticists, systematists, and paleontologists. Not only were these disparate phenomena—these genes, species, and phyla—all part of the evolutionary picture, but the early influential architects of the synthesis saw that nothing said about any one particular phenomenon could be inconsistent with what was known about the others (as Simpson explicitly recognized in 1944 for genes and fossils). Perhaps the best analogy would be the dictum that no scientific formulation about a system can contravene more general natural laws. Mechanics of heredity, development, speciation, and the like must at the very least be consistent with, say, the Laws of Thermodynamics—if not actually reducible to these laws (as Wiley and Brooks suggested in 1982 for some evolutionary phenomena).

Thus, consistency was a desideratum in the early writings of the synthesis, as has recently been stressed by Gould (1980b, 1982b), Mayr (1980a, 1982), and myself (1982a). Mayr's often-quoted "one-two" punch summary of the synthesis, while obviously an oversimplification (but see chapter 4 herein), nonetheless nicely epitomizes the general structure of the main argument of the synthesis:

The term "evolutionary synthesis" was introduced by Julian Huxley in *Evolution: The Modern Synthesis* (1942) to designate the general acceptance of two conclusions: gradual evolution can be explained in terms of small genetic changes ("mutations") and recombination, and the ordering of this variation by natural selection; and the observed evolutionary phenomena, particularly macroevolutionary processes and speciation, can be explained in a manner that is consistent with the known genetic mechanisms. (Mayr 1980a, p. 1)

In other words, natural selection, working on a groundmass of variation (produced ultimately by mutation), is the major deterministic cause of organic change—and all patterns of biological change recorded by geneticists, systematists and paleontologists are consistent not merely with the idea that selection occurs, but with the stronger notion that such change is, at base, the direct product of a selection process.

The synthesis has recently been criticized as being strongly reductive (e.g., Eldredge 1982a, 1982b; Gould 1982b), in part a reaction to such formulations as Bock's (1979) paper, which explicitly seeks to reduce macroevolution to microevolution. But recent advances in molecular biology have spawned entirely novel theories of genomic change—explicit theories of evolutionary change at the molecular level which are not in the least foreshadowed by the older theories well established in what Dobzhansky (1937a) calls "physiological genetics" and the genetics of populations. Of course, we must all continue to hearken to the consistency argument so clearly articulated by the founders of the synthesis. But such notions as Dover's (and those of his colleagues; see especially Dover 1982) "molecular drive" are additional theories specifically addressed to a level of biological organization (internal structure of the genome within an organism) not itself addressed by the synthesis—for the good and sufficient reason that virtually nothing of the molecular nature of heredity was known in the 1930s and 1940s. Thus, simply calling the synthesis reductionist is true to its spirit of looking up from the atomistic, particulate gene—on up through populations, species, and phyla—but it is a rather inadequate description of the present state of affairs.

It is more revealing to see that the synthesis was addressed primarily to but a few basic *levels* of the evolutionary process—through simple ignorance of lower-level phenomena and a perhaps genuine desire to reduce higher-level phenomena. As we shall shortly see, Dobzhansky had a definite hierarchical view of the evolutionary process. Genetic processes within individual organisms (especially mutation as a counterforce to accurate hereditary transmission) work at one level. But when one considers the fate of among-organism allelic variation (i.e., within populations), the "rules" (his term) of the game change and involve, inter alia, selection, drift, and such factors as population structure and size. Species represented a still higher rung. Thus, a triad of genes, organisms and populations (with species as a fourth level) forms the basic structure of Dobzhansky's (1937a) view of evolution. His was no simple theory. And reductionism is also too simple a characterization if what is meant is that all the early architects saw nothing but the genetics of populations as the whole of the evolutionary process.

I have devoted chapters 2 and 3 of this book to a careful analysis and exegesis of the four main early works that led to the synthesis: Dobzhansky (1937a, 1941), Mayr (1942), and Simpson (1944). The rationale for choosing these four books is simple enough: taken together they determine much of the content and all of the structure of evolutionary theory—the theory that is still very much paramount today. After a brief comparative overview in chapter 4, I enumerate the major focal shifts in both structure and content of the synthesis since the books were written—and I come up with only three shifts that qualify.

Thus, the first section of this book (chapters 2 and 3) is mainly about four other books. The second section (chapters 4 and 5) summarily characterizes what the synthesis is, how it has indeed changed from the early days into its modern form, and how some of the challenges that have materialized since those early days (challenges that for the most part reflect internal difficulties and inconsistencies within the synthesis since its inception) require a fairly basic restructuring of evolutionary theory. Finally, in chapters 6 and 7, I offer my version of what that restructured theory should look like. Throughout the book I will conclude that the synthesis is not so much incorrect as incomplete. For the most part, the processes of evolution are still considered (by paleontologists as well as geneticists) to remain almost wholly within the purview of geneticists. Now, this view would be fine if it provided an accurate description of how the real world is structured and how evolution functions. But it leads to a rather lopsided and distorted theory if the world is not organized in such a fashion—that is, if there are processes in nature that are not suited to analysis in laboratory populations of house mice and fruit flies. Simpson made a similar, if more subtle, point when he suggested that the scale of events in evolutionary time gives us an utterly different perspective on how "evolutionary determinants" (which he saw as largely phenomena of genetics) really work together in the evolutionary process (Simpson 1944, p. xvii; quoted in full in chapter 3 of this book).

A major effect of the synthesis, in my estimation, has been the removal of most biological disciplines, except genetics, from the active ranks of investigation into the evolutionary process. Once again, this is not necessarily bad—if, that is, the focus of the synthesis on genetic change is really all we need to understand all aspects of evolutionary stasis and change. But this epistemological aspect of the synthesis (an aspect that otherwise does not concern me greatly in this book) reflects a deeper and far more interesting problem in contemporary evolutionary theory: the matter of ontology.

In a nutshell, the synthesis limits its attention to only a few of the biological entities that seem to me to exist in the world and to be involved in the evolutionary process. Genes (in a premolecular phase of understanding of course), organisms, demes (to some degree), and species are explicitly addressed in the writings of the synthesis; monophyletic taxa are but dimly perceived: and ecological entities (populations, communities, and regional biotas) are not even addressed. Species are especially crucial; I will show that the view of species in the synthesis reveals a variable ontology that is context-dependent. Species are quite real to Dobzhansky and Mayr when considered at any point in time and particularly when sympatric with other closely related species. Through time, though, species are classes or, at best, classlike entities, viewed in a manner that is literally required if we see evolution as primarily the adaptive transformation of phenotypic attributes through time. Viewing species (and other entities usually treated as classes by the synthesis) as individuals—following the lead of Ghiselin (1974a) and Hull (1976)—enlarges the range of entities that take an active part in the evolutionary process. What we need, it seems, is a revised ontology (Hull 1980) of evolutionary entities. I address this topic at length in chapters 4 through 7 of this book.

This revised ontology, I will argue, automatically forces us to consider an alternative approach to the very structure of evolutionary theory— simply because it presents us with an alternative description of the organization of biological nature. That structure is *hierarchical*. Genes, organisms, demes, species, and monophyletic taxa form one nested hierarchical system of individuals that is concerned with the development, retention, and modification of *information* ensconced, at base, in the genome. But there is at the same time a parallel hierarchy of nested *ecological* individuals—proteins, organisms, populations, communities, and regional biotal systems, that reflects the *economic* organization and integration of living systems. The processes within each of these two process hierarchies, plus the interactions between the two hierarchies, seems to me to produce the events and patterns that we call evolution.

But mere recognition that the biological world is hierarchically organized constitutes no compelling reason to conclude that the theory of adaptive change that essentially *is* the synthesis is either inaccurate or incomplete. We must further show that there are questions left unanswered by the synthesis—problems not handled comfortably by the prevailing paradigm (the sort of problems beloved of philosophers and historians of science) or simply entire sets of questions not even considered for some reason under that paradigm. Examples of the former (such as those given by Hull 1980, p. 311) include the prevalence of sex, the presence of relatively large amounts of genetic variation within most populations, and selection at higher levels. Examples of the latter include virtually all species-level phenomena in the history of life, including trends, "radiations," and the like; cross-genealogical extinction events (of varying magnitude) that degrade ecological systems; and the likewise cross-genealogical patterns of proliferation following such episodes.

From the perspective of a paleontologist–systematist, the synthesis fails at base because its very description of nature at and above the species level is faulty. Ironically, we need only to examine the notion of natural selection to discover the roots of the problem. (I say "ironically" because there is manifestly nothing wrong with the concept of natural selection itself; differential reproductive success among sexually reproducing organisms seems to be a very well-established natural process.) Following Hull (1980), we see that there are two ingredients involved in any form of selection at any level. For selection to occur, there must be what Hull calls "replicators" and "interactors." In other words, we need some entity that makes more of itself (i.e. with reasonable accuracy) *and* engages in some manner (typically involving matter–energy transfer) with its milieu. Organisms, of course, fulfill these requirements, and thus natural selection flows as a logical necessity. Organisms both reproduce (with genes supplying replicative fidelity) and interact in nature, producing the well-known, virtually inevitable bias in reproductive success that has come to be the modern conception of natural selection.

Hull's two ingredients of selection are actually the twin basic elements that give us "life": a replicative system of "nucleic" acids and an energetics system (now arising from replicative templates) that both houses the replicating molecules and catalyzes their replication. The trend in recent years, seen especially clearly in the works of Dawkins (1976, 1982), has been to see evolution fundamentally from the gene's perspective: evolution is a simple outcome of genic strategies to leave more copies of themselves. In many ways, as I shall discuss, Dawkins's "selfish gene" is a compelling metaphor. Yet such a conception in effect claims that the information of the genome somehow takes primacy over the energetics system in which it is housed and upon which it depends for its very function. Is an organism just a genome's way of making another genome? Or are genomes just an organism's way of (1) staying alive (somatic genome) and (2) making more organisms (germ-line genome)? Both positions can be defended, of course; in general, the answer to both questions is yes. But it is important to realize that it is logically impossible to establish the primacy of one view over the other.

August Weismann, in the late nineteenth century, taught us to distinguish between the soma and the germ line. Again, such a distinction starkly recognizes the dual nature of life, the fact that life involves both reproduction (information conservation and transmission) and economics (matter–energy transfer). In metazoans and multicellular plants, the germ lines and soma remain, as a rule, recognizably distinct, and Weismann's doctrine (that the soma develops from parental germ-line products, but what happens to the soma cannot affect germ-line cells) has remained intact, despite recent claims to the contrary. (In prokaryotes and simpler forms of eukaryotes, of course, the distinction between soma and germ-line is blurred.) But the mere fact that complex organisms are both a soma and a germ line (the fact that renders natural selection a logical necessity) obscures an equally fundamental point about biological organization above the organism level. If we take the position that there are biological entities—individuals—at levels above the organism, we note right away that the dual fundamental characteristics of life remain but are arrayed in a somewhat different manner from what we are used to seeing in organisms.

If the distinction between the two functions (economics and reproduction) is anatomically vague in prokaryotes and simple eukaryotes and fairly clear albeit housed within a single entity in more complex organisms, the two functions are distributed within separate, noncomparable entities in biotic nature above the organism level. Put another way, above the organism level economics (matter–energy transfer) and reproduction (information conservation and transfer) become segregated into systems far more formally discrete than Weismann's soma–germ line distinction. Above organisms, biologists historically see *either* demes, species, and monophyletic taxa of higher categorical rank *or* populations, communities (or local ecosystems), and regional biotas. Thus, in the description of biotic nature that I favor later in this book, I follow convention and see the biotic world—meaning all of life at any one moment—distributed simultaneously into two hierarchical systems: the genealogical hierarchy (with its genes, chromosomes, organisms, demes, species, and monophyletic taxa) and the ecological hierarchy (with its proteins, the somatic hierarchy of organisms, populations, communities, and larger ecological units).

What possible difference could this description make to understanding evolutionary theory? For one thing, consider the problem of selection at levels both lower and higher than the among-organism, within-deme level. If my arguments in chapters 5 through 7 are correct, and accepting the validity of Hull's twin necessary conditions for selection, no entity above the organism level can be selected (at least in a sense consistent with use of the term *natural selection*). I hasten to add that species and other entities may—indeed commonly are—"sorted" (see also Vrba and Eldredge 1984). But note, too, that the assertion of the synthesis—that all large-scale evolutionary phenomena essentially fall out of the adaptation-through-natural-selection paradigm—also treats species and monophyletic taxa as economic entities, perched as they are on "adaptive peaks," "ranges" and so forth.

Thus, a revised ontology simply expands the range of possible dynamic sorting processes of various levels of biotic entities. The synthesis deals with species and monophyletic taxa by extrapolating what goes on among organisms within populations (seeing organisms properly as both economic and genealogic entities) and viewing higher taxa as nothing but classlike strings of sexually reproducing organisms moving through time. In the synthesis, species and higher taxa are primarily economic entities— to the extent that they are seen as entities at all. In contrast, I shall argue that species are (1) individuals (spatiotemporally bounded historical entities) and (2) genealogic entities (i.e., informational, *not* economic in nature). The extrapolationist model of macroevolution based on a simple paradigm of neo-Darwinian processes cannot possibly stand as a truly accurate description of nature. A hierarchical theory based on a clarified ontology of large-scale biological entities yields a much more realistic approach to macroevolution, specifically by formally integrating issues of differential births and deaths of these various sorts of upper-level biotic entities into mainstream evolutionary theory.

Later chapters present my formal arguments on the "individuality" of such entities as chromosomes, species, communities, and so forth. There is, I believe, a sort of glue that gives cohesion to each entity. In the genealogical system, that glue generally seems to be the simple, ongoing more-making of next-lower-level individuals. Organisms make more organisms to keep species going (or demes, if admitted to the system); species make more species to keep monophyletic taxa going. Similarly, economic interactions among organisms keep populations going, and the among-population dynamics supply cohesion to local ecosystems.

Hence my earlier statement: it is processes within each hierarchy plus the interactions between elements of the two hierarchies that give us this thing we call evolution. If evolution is stasis and change in genetic information through time, that information is (minimally) packaged (1) in each genome, (2) among organisms within demes, (3) among demes within species, and (4) among species within monophyletic taxa. Anything contributing to biases in the births and deaths of any of these entities ("sorting" according to Vrba and Eldredge 1984) is thus material to an understanding of the evolutionary process. But just as real in the biological arena is the moment-by-moment economic organization of life. These units— informational and economic entities—may now be segregated (i.e., above the organism level), but the interactions between entities of the two hierarchies mimic selection very well. In particular, biases in births and deaths of genealogical units (demes, species, monophyletic taxa) arise from the fates of ecological entities, particularly the fates of organisms integrated into the economy of nature. The genealogical hierarchy supplies the players in the economic game of life; their fate determines the future representation of information in the genealogic hierarchy. The dual, basic features of life—replication and interaction in Hull's terminology, or simply reproduction and economics (so clearly seen in natural selection)—are the essence of the entire organization of life at all levels and are also essential to a more accurate description of nature and a firmer grasp of the evolutionary process. A more accurate description of nature—one that follows the general lines I have elaborated here—does indeed offer a greater range of explanations of, or simply ways of looking at, the problems that Hull (1980) sees as residual conundra for synthesis-inspired, organism-centered evolutionary theory. And on the very face of it, such a description admits that higher-level ecological and genealogical phenomena (which Teggart saw in 1925 as the actual events in the history of life) become part of what a theory of the evolutionary process seeks to explain.

To return to epistemology for a moment, hierarchies also afford alternative means of grappling with complexity. If the overall gestalt of the modern synthesis is elegant simplicity, its drawback (or so it seems to me) is that it has left too much out. When approaching complexity, we need to simplify if only to encompass a phenomenon, a requisite first step toward understanding what goes on in any system.

Hierarchies at first glance seem to make matters much more difficult, providing no help at all in managing complexity because they are themselves so complex. But hierarchies actually deal with complexity by teasing it apart; it is as if hierarchies are more honest in their simple recognition that a system is complex than is an approach that seeks unity in characterizing the system in simple terms—such as in the description of evolution provided by later versions of the synthesis. The synthesis says that all evolutionary phenomena are to be understood with reference to a relatively noncomplex set of within-population, among-organism dynamics that have direct effects on the representation of genetic information in each successive generation. Epistemologically, this is a nightmare for molecular biologists, embryologists and paleontologists alike, for, except in rare circumstances, there can be no direct application of theory to data in the manner considered de rigueur in the canons of science.

Hierarchy ameliorates this situation considerably. In the description of the genealogical and ecological hierarchies, the observations and experiences of biologists working on the diverse kinds of biological individuals automatically become more relevant to an understanding of the nature of the evolutionary process than the synthesis allows to anyone except geneticists. In the synthesis, those of us who are not geneticists must be content with the fact that the phenomena we record are simply the products of genetic processes that we, alas, cannot directly study ourselves.

The hierarchical system outlined later in this book provides a framework for understanding evolution as the more complex affair that it is, more so than is depicted by the synthesis. Processes are seen to be inherent in each level within both hierarchies, and there are, as well, patterns of upward and downward causation within each hierarchy. And there are interactions between the two hierarchies. In insisting that the evolutionary process is more complex than admitted by the synthesis, however, by no means do I revert to the presynthesis no-holds-barred competition among disciplines, each vying for primacy in describing one part of the complex whole as though it somehow constituted a description of the entire system—the "blind men and the elephant" routine. That was what presynthesis evolutionary biology seems to have been like. The synthesis vastly improved the situation by focusing on one large, very important part of the elephant—roughly two-thirds of one of the hierarchies. Some of its proponents persist in thinking that we have thereby gotten more than a rough outline of the entire elephant; that, however, is open to dispute.

As will soon become clear, the early architects of the synthesis, especially Dobzhansky (1937a, 1941), were very conscious of hierarchy. My elaboration of hierarchy theory in the later stages of this book is really nothing more than a return to this early perspective, coupled with an enlarged and (I hope) improved ontology of biological entities that leads to a fuller description of evolutionary events, entities, and processes. Indeed, this return sheds some new light on some very old problems, such as the retention of variation within populations and the prevalence of sex—problems that crop up in early synthesis writings and still plague contemporary theorists working under the basic paradigm of the synthesis.

It is undoubtedly naive to hope that the epistemological ramifications— where such disparate fields as ecology, paleontology, systematics, and molecular and developmental biology may at last be truly functionally integrated into evolutionary biology—will automatically follow as a result of thinking about evolution in hierarchical terms. But it is worth a try.

# 2

# Genes and
# the Evolutionary Synthesis

"What is still lacking is a critical analysis of the writings of the architects of the synthesis."
(Mayr 1982, p. 568)

My intention in this and the following chapter is simply to clarify what a few of the better known architects of the synthesis actually had to say about how the entire spectrum of genes through phyla really fits together. I have chosen books rather than a potpourri of articles from the (possibly) more "technical" scientific literature, for precisely this reason: it is in the books that we find the coherent, integrated statements. And each of the four books singled out for particularly close analysis—*Genetics and the Origin of Species* (Dobzhansky 1937a); *Genetics and the Origin of Species* second edition (Dobzhansky 1941); *Systematics and the Origin of Species* (Mayr 1942); and *Tempo and Mode in Evolution* (Simpson 1944)—is a truly coherent, though not necessarily smoothly linear, argument. In each, some parts seem more vital to the flow of argument than others, but it is clear to the reader from the outset that each is a complete book and not a disjointed melange of unrelated ideas.

Then, too, the contents of each author's papers are to a great extent reflected in the respective books. It is apparent, for example, that Dobzhansky was publishing a variety of papers in the 1930s (many were in his "Genetics of Natural Populations" series republished in a single volume in 1981 by Lewontin et al.) which he liberally drew upon for illustrative material in *Genetics and the Origin of Species*. The other sorts of papers Dobzhansky was publishing seem to be byproducts of his thinking and research for the various editions of his book. Examples are his papers on species definitions (1935) and isolating mechanisms (1937b) and his theory of the origin of isolating mechanisms by reinforcement (1940).

The choice of the four particular books for special treatment in this volume needs further comment. I have omitted such nonconformist works as Robson and Richards (1936), Willis (1940), and Goldschmidt (1940) precisely because they have been almost universally considered as falling outside the limits of the synthesis. The main historical effect of such books (especially Goldschmidt's) seems to have been as irritants. Dobzhansky's second edition (1941), Mayr (1942) (as Mayr himself has said, cf. Hapgood 1984), and Simpson (1944) all take up the cudgels against Gold-schmidt's theme of discontinuity and his thesis that within-species pro-cesses of change (especially selection) are not the stuff of among-species differences. Missing too is Huxley's *Evolution: The Modern Synthesis* (1942)—within the confines of the emerging synthesis to the point of lending it a name. I also omit Rensch (1947, 1960), Simpson (1949, 1953), Stebbins (1950), and other "classics" of the synthesis that are on nearly everyone's list. My reasons are simple: in addition to the agreed-upon primary status of these four books (see, for example, the continual reference to the three authors and their books throughout the 1980 vol-ume on the history of the synthesis edited by Mayr and Provine), my pur-poses are not historical. I am not interested in tracing the roots of the synthesis, its historical antecedents, or the sequences of its development. I am not concerned with adducing the roles played by assorted figures from various disciplines (such as cytogenetics, population genetics, sys-tematics, paleontology, etc., as Mayr—1980a and 1982—has been espe-cially concerned to do). My own opinion on such historical issues, for what it is worth, is that talented individuals of admittedly differing background and training were the ones most responsible for forging the consistency statement that is the synthesis.

In what follows, some light no doubt will be shed on historical aspects of the development of the synthesis. I was totally unprepared, for exam-ple, to find that, above the organism level, Dobzhansky's view of the structure of species and the interplay of evolutionary processes (his "dynamics") was based so completely on the view Wright had developed in the early 1930s (especially Wright 1932). And some ideas added to his second edition (Dobzhansky 1941) clearly anticipate arguments expressed in Mayr (1942) and especially Simpson (1944). This is interest-ing, to be sure, but by no means the quest of my research. I am interested in what was said.

Dobzhansky, Mayr, and Simpson simply produced the earliest clear statements of what has become the synthesis. At least in the United States, these were the books that everyone read just to learn about the latest thinking in evolution (just as it was Dobzhansky 1951, Simpson 1953 and 1961, and Mayr 1963 that I got to know intimately in graduate school, having to discover for myself the virtues of the earlier works). And, per-haps most importantly, each of the three had a different perspective. Dob-zhansky, though trained as a naturalist (see, for example, Levene et al. 1970; Gould 1982a), comes at the problem explicitly from the vantage point of the gene; he builds his argument up from the gene to the spe-cies—and beyond. Mayr's point of departure, in contrast, is species, and he looks at genes and populations below, and taxa of higher rank above, from just this perspective. Simpson, I will argue, aimed his entire book at the explanation of the origin of taxa of higher rank, and his views of genes and species are likewise affected by his own interests and background as a paleontologist.

Why do we need a critical assessment of these works? Simply, my argument that the ontology of the synthesis is incomplete and partly incorrect demands documentation. The structure of evolutionary thinking that emerges from close reading of these books has remained substantially unaltered since they were written—a rather serious charge that itself requires examination. Just what this structure is, and what its underlying ontological basis might be, is best revealed by a careful dissection of these works.

What follows is a critical assessment of the arguments of each of the four books, taken in order of publication, with commentary on Dobzhansky's second edition, of course, restricted to the differences that show up between the two editions.[1] I shall then summarize briefly by comparing the basic evolutionary theories of the four books and their (to some extent disparate) views on the ontological status of such biological entities as genes, species, and phyla. This summary, in turn, provides the basis for a consideration of later works by these and other authors simply to round out a characterization of the structure and content of the modern synthesis. We will then be in a position to evaluate Stebbins and Ayala's (1981) claim that the synthesis is broadly enough based to subsume more recent developments in evolutionary theory.


## GENETICS AND THE ORIGIN OF SPECIES

In the preface to *Genetics of the Evolutionary Process*, Dobzhansky (1970) admits his original intention simply to dub that work the fourth edition of *Genetics and the Origin of Species*. Cooler heads prevailed.[2] Both the organization and the content of the 1970 book are rather different from the third edition of *Genetics and the Origin of Species* (1951), and it was time at last to abandon that old title. But the fourteen years that separated the first and third editions saw a change in emphasis and position that is partly masked by Dobzhansky's conservative retention of chapter titles, basic organization, and much of the original verbiage, diagrams, and tables. Yet the additions, deletions, and rearrangements make the second edition (1941) rather different in important ways from the original version, and the 1951 book in turn is somewhat different from the 1941 second edition.

Dobzhansky (1937a) uses his first chapter effectively. In it he reveals his basic approach, his fundamental views about what nature is and how we find out about it. He develops his hierarchical view of the evolutionary process, and in so doing (in barest outline only) he exposes the rudiments of his own theory of that process. He speaks of evolutionary "statics" and "dynamics." Statics "treats of the forces producing a motion and the equilibrium of these forces," while dynamics "deals with the motion itself and the action of forces producing it"(p. 12). This is his classification of the variables in genetics, such as mutations and chromosomal changes (the statics) versus the population-level processes of drift and selection (the dynamics) that take genetic variation and mold it into evolutionary change. He promises and duly delivers a book organized according to this classification: statics are treated in chapters 2 through 4, while dynamics fill the rest of the book.

Dobzhansky is, above all, a realist and a pragmatist. To him, there are two facts about the natural world that, together, form the twin problems for evolutionary theory to explain: *diversity* and *discontinuity* (a theme explicitly adopted by Mayr in 1942; see chapter 3 of this book). The whole point of Dobzhansky's chapters 2, 3, and 4 is to establish the genetic basis of this diversity (meaning phenotypic differences, which Dobzhansky, interestingly, explicitly states he is not addressing per se; see his p. 7). Genes—particulate, atomistic entities of unknown structure and composition that lie in linear arrays on chromosomes (his p. 74)—underlie the differences among organisms, populations, species, and on up through phyla. Thus, the diversity is essentially continuous. The origin of this diversity is ultimately mutation. The origin of *dis*continuity, on the other hand, is a major focus of the later sections of Dobzhansky's book and remains an unsatisfactorily resolved problem that he attacks again, with renewed vigor, in his second edition (1941). There are two sources of discontinuity: the particulate nature of genes and the quantal aspects of mutations on one level, and Dobzhansky's famed "isolating mechanisms" which separate species (about which more later).

The opening chapter also accurately conveys Dobzhansky's notion of the essence of evolution and the role played by genetics in the science of evolutionary biology. Though he takes pains to say (p. 8) that "genetics as a discipline is not synonymous with the evolution theory, nor is the evolution theory synonymous with any subdivision of genetics," nonetheless at various places the synonymy is very nearly made:

Since evolution is a change in the genetic composition of populations, the mechanisms of evolution constitute problems of population genetics (p. 11)

Here is another fairly bald statement:

Evolution is essentially a modification of this equilibrium. (p. 124, in reference to "Hardy's formula," known more familiarly today as the Hardy-Weinberg equilibrium)

Now, it is certainly easy to take selected quotes—even preserving their context—and overlook intent and the fact that an author may reasonably be expected not to be utterly consistent throughout. Certainly no one would read Dobzhansky's book and conclude that Dobzhansky thought that all evolution really amounts to is change in gene frequencies. It is fair to say, however, that he sees diversity as essentially boiling down, not merely to changes of allelic frequencies in populations, but right on down to the origin of mutations that have relatively small effects and are, if not beneficial or neutral, perhaps only slightly deleterious.

That Dobzhansky sees no simple reduction of evolution to the principles of genetics comes not least from his appreciation that genetics involves several subdisciplines addressed to rather different sorts of phenomena; hence his explicit hierarchical viewpoint already mentioned. The extended quote that follows provides a useful summary, in Dobzhansky's own words, of his evolutionary theory as of 1937:

In bare outline, the mechanisms of evolution as seen by a geneticist appear as follows. Gene changes, mutations, are the most obvious source of evolutionary changes and of diversity in general. Next come the changes of a grosser mechanical kind involving rearrangements of the genic materials within the chromosomes. It seems probable at present that such rearrangements may at least occasionally entail changes in the functioning of the genes themselves (position effects), since the effects of a gene on development are determined not only by the structure of that gene itself but also by that of its neighbors. Finally, reduplications and losses of whole chromosome sets (polyploidy) are important as evolutionary forces, especially among some plants.

Mutations and chromosomal changes arise in every sufficiently studied organism with a certain finite frequency, and thus constantly and unremittingly supply the raw materials for evolution. But evolution involves something more than origin of mutations. Mutations and chromosomal changes are only the first stage, or level, of the evolutionary process, governed entirely by the laws of the physiology of individuals. Once produced, mutations are injected in the genetic composition of the population, where their further fate is determined by the dynamic regularities of the physiology of populations. A mutation may be lost or increased in frequency in generations immediately following its origin, and this (in the case of recessive mutations) without regard to the beneficial or deleterious effects of the mutation. [As we shall see below, this is a reference to what Dobzhansky here calls "scattering of the variability" and later (1941) calls by the familar term "genetic drift."] The influences of selection, migration, and geographical isolation then mold the genetic structure of populations into new shapes, in conformity with the secular environment and the ecology, especially the breeding habits, of the species. This is the second level of the evolutionary process, on which the impact of the environment produces historical changes in the living population.

Finally, the third level is a realm of fixation of the diversity already attained on the preceding two levels. Races and species as discrete arrays of individuals may exist only so long as the genetic structures of their populations are preserved distinct by some mechanisms which prevent their interbreeding. Unlimited interbreeding of two or more initially different populations unavoidably results in an exchange of genes between them and a consequent fusion of the once distinct groups into a single greatly variable array. A number of mechanisms encountered in nature (ecological isolation, sexual isolation, hybrid sterility, and others) guard against such a fusion of the discrete arrays and the consequent decay of discontinuous variability. The origin and functioning of the isolating mechanisms constitute one of the most important problems of the genetics of populations." (Dobzhansky 1937a, pp. 12–14)

This quote provides an accurate precis of both the generalities of Dobzhansky's theory and the ensuing structure of his book. Before getting to his extended argument, there is one final issue broached in his introductory chapter: macroevolution.

Rather than stating as Stebbins and Ayala (1981) have recently done on epistemological grounds, that it might serve as well to entertain the notion that micro- and macroevolution are, if not independent, then at least not utterly coterminous, Dobzhansky says that "we are compelled at the present level of knowledge reluctantly to put a sign of equality between the mechanisms of macro- and micro-evolution, and, proceeding on this assumption, to push our investigations as far ahead as this working hypothesis will permit" (Dobzhansky 1937a, p. 12).

Thus, for lack of something better, we must be content with an integrated theory of evolution that begins and ends with the parameters of genetics as the sole determinants (Simpson's 1944 phrase for a slightly different set of Dobzhansky's statics and dynamics) of the entire evolutionary process. It is a brave task Dobzhansky sets for himself for the remainder of the book—and he does an admirable job. Later he tinkered with the few points of his original 1937 treatment that he felt really didn't do the job, with the result that the second edition (1941) is a bit smoother and a somewhat more complete (and certainly more definite) statement. But the remainder of the original edition is no slouch either.

### Evolutionary Statics

Summarizing his chapters 2 through 4 at the outset of chapter 5, Dobzhansky tells us (p. 118) that "it is now clear that gene mutations and structural and numerical chromosome changes are the principal sources of variation." This is precisely the argument of the preceding 102 pages, packed as they are with laboratory experimental results and observations of similar sorts of phenomena in the wild. The hierarchy Dobzhansky sees in the natural world has more to do with his dynamics than with his statics: there is a difference between the processes of mutation and chromosomal rearrangement on the one hand and their fates when subjected to drift and selection within populations on the other hand. There are, to be sure, small-scale effects (mutations) and larger-scale ones. But the point is clear: there is a genetic basis to almost all the sorts of differences one sees, within a small laboratory population, between populations, or between species. And though Dobzhansky addresses no section to taxa of rank higher than species, the otherwise inexplicable final section of chapter 3—"Maternal Effects and the Non-Genic Inheritance"—asks (and answers) whether "natural variability" has any cause other than "gene variation." Hybrids between organisms belonging to different taxa of relatively high rank (families, etc.) being rather rare, it had been suggested by "a number of biologists, with some very eminent names included," that "genes determine only the characteristics of the lowest systematic categories (races, species, genera), while those of the higher categories[3] are not genic" (p. 68). So Dobzhansky duly and patiently reviews the literature on cytoplasmic inheritance, only to conclude that it is "so rare relative to genic inheritance that in the general course of evolution the former can hardly play more than a very subordinate role" (p. 72). Thus, without being stated explicitly, the very "genic variability" underlying the differences between races, species, and genera underlie the differences on up the Linnaean hierarchy. In terms of Dobzhansky's statics, diversity is fully continuous at base, and his hierarchy of entities—genes, organisms, populations, and species—is fully transitive (in the sense of Eldredge and Salthe 1984) insofar as the origin of that diversity is concerned.

When Dobzhansky turns to the notion of race, he explicitly denies his concern with morphology per se—though, of course, all the evidence pointing to the genetic basis of inheritance and the nature of mutation is based on the occurrence of morphological attributes. Up to this point in his text, he has been concerned with the distribution of mutations and allelomorphs in general within populations and species. He now switches to a different ontological question: What is a race? His answer is clear and totally consistent with his later definition of species: a race is not a particularly clear cut entity. Dobzhansky tells us (p. 62) that "the fundamental units of racial variability are populations and genes, not the complexes of characters which connote in the popular mind a racial distinction." In a striking passage, he maintains that a race is "not a static entity but a process" and goes on to say that a race has stages of development:

If the differentiation is allowed to proceed unimpeded, most or all of the individuals of one race may come to possess certain genes which those of the other race do not. Finally, mechanisms preventing interbreeding of races may develop, splitting what used to be a single collective genotype into two or more separate ones. When such mechanisms have developed and the prevention of interbreeding is more or less complete, we are dealing with separate species. A race becomes more and more of a "concrete entity" as this process goes on; what is essential about races is not their state of being but that of becoming. But when the separation of races is complete, we are dealing with races no longer, for what have emerged are separate species. (Dobzhansky 1937a, pp. 62–63)

Races are merely stages of development—Dobzhansky's "process"— which he could not have meant literally as they are treated in this passage more like labile entities than the pure classes he implies them to be. Species here, by implication, are fully concrete entities, the final stage of racial differentiation. But later in the book (p. 312), species are also seen as "a stage in a process, not a static unit," a notion later criticized explicitly by Mayr (1942). It is as though, viewed from the particulate gene, Dobzhansky sees an intergradation of constellations of organisms (themselves, of course, particulate, but not expressly dwelt upon as such in his book) from small populations up to races and on to species. And when the latter plateau is reached, we once again confront discontinuity. (Later on in the book, "microgeographic races," Wright's "colonies," or semi-isolated subdivisions of a species are added to the list of such constellations of organisms, but with a very different end in mind.) Again, the simple point—but one utterly crucial to any blend of neo-Darwinian dynamics with physiological genetics—is the convincing demonstration of mutation as the underlying source of all morphological variation in nature.

Dobzhansky concludes his discussion of evolutionary statics with a survey of chromosomal variation in the lab and in the wild. After judicial review of lab and field evidence, Dobzhansky says (p. 114) that "there can be little doubt that chromosomal changes are one of the mainsprings of evolution." And, as he turns to address the dynamics of the evolutionary process, the position of genetics within evolutionary biology is blindingly obvious because all those features in which organisms have come to differ—all that diversity out there—has a genetic basis, with mutation as its ultimate source. The final two-thirds of Dobzhansky's book looks at just how those genes are apportioned within aggregates of organisms. Much of what he has to say on this subject will come as a surprise to those familiar with later versions of the synthesis, including some of Dobzhansky's own later writings.

## Evolutionary Dynamics

Plasticity—not putting all your evolutionary eggs in one basket but retaining instead a sufficient reservoir of variability to handle and respond to future environmental change—runs as a consistent theme throughout Dobzhansky's writings right until the end of his career. The problem is to maximize adaptation without specializing too much; overspecialization reduces the pool of genetic variability available to a population or species to meet those exigencies of changing environments. Yet selection can only work to maximize the short-term adaptations of organisms and has no eyes for the future. This conflict, this tension between the short-term benefits of adaptive maximization and the long-term benefits of survival accrued to species that retain variability (at the expense of honing their short-term adaptations), provides the key to understanding Dobzhansky's rather elaborate vision of evolutionary dynamics in the first edition of *Genetics and the Origin of Species*. It remains a difficult problem in evolutionary theory and forms a crucial aspect of the whole issue of hierarchy in evolution.

Dobzhansky begins the characterization of evolution's dynamics in his chapter 5, "Variation in Natural Populations." After summarizing the foregoing "continuity in the genetic basis of diversity" argument, he asserts strongly the first of the two main points of the chapter: the fate of genetic variability within populations is determined by a set of rules utterly different from those that obtain for the origin of genetic variation within organisms. His position here is notably hierarchical: the fate of hereditary variation obeys the "rules of the physiology of populations, not those of the physiology of individuals" (p. 120).

Dobzhansky also remarks at the outset of chapter 5 that

in nature we do not find a single greatly variable population of living beings which becomes more and more variable as time goes on; instead, the organic world is segregated into more than a million separate species, each of which possesses its own limited supply of variability which it does not share with the others. (Dobzhansky 1937a, p. 119)

How is that variability structured, partitioned, allocated, and changed? Plunging into the work of theoretical (i.e., mathematical) population genetics, Dobzhansky first discusses the mathematical consequences for the fate of variability if one adopts the particulate theory of heredity instead of the notion of blending inheritance. He turns next to genetic equilibria and the effects of mutation thereon—the first appearance of simple mathematics in the book. It is here that we encounter the second main point of this chapter: the tension between long-term plasticity and the short-term benefits of close adaptation:

Evolutionary plasticity can be purchased only at the ruthlessly dear price of continuously sacrificing some individuals to death from unfavorable mutations. Bemoaning this imperfection of nature has, however, no place in a scientific treatment of this subject. (Dobzhansky 1937a, p. 127)

Tackling the problem, Dobzhansky completes the chapter with a review of theoretical and observational investigations of variation within populations. First he examines what he calls "scattering of the variability." There is an agent (we know it now as genetic drift) that, in addition to selection, acts against mutation to reduce genetic variability. Drift (the expression "genetic drift" per se does not appear in Dobzhansky's text until his second edition) relies on mere chance to reduce variation. Here we contact for the first time in depth the ideas of Sewall Wright, whose concepts, published in the early 1930s (especially Wright 1932), come to dominate Dobzhansky's views on the organization of nature and the actions of evolutionary dynamics throughout the remainder of his book.

The balance of the chapter, in effect, picks up and explores three interrelated notions of Wright: (1) that the process of sexual reproduction alone can lead to frequency changes quite apart from selection (i.e. the action of genetic drift), (2) that this process is most effective in populations of "intermediate" size, and (3) that species tend to be "subdivided into numerous isolated colonies of different size, with the exchange of individuals between the colonies prevented by some natural barriers or other agents. As we shall attempt to show below, such a situation is by no means imaginary; on the contrary, it is very frequently encountered in nature" (Dobzhansky 1937a, p. 133, and found throughout the remainder of the book). Indeed, this view of the internal structure of species becomes the very essence of Dobzhansky's vision of species in the wild and entirely underlies his later scenarios of evolutionary dynamics.

To support his contention that species indeed tend to be organized into semi-isolated colonies, Dobzhansky briefly surveys the biological literature, first reviewing Gulick's and Crampton's studies of Pacific land snails, citing with evident approval (p. 136) their reluctance to ascribe the differences between local races (in different drainage valleys) to natural selection. After more examples, he turns to the literature on population size, acknowledging that the data are meager but that at least they "tend to show that the effective population sizes may prove to be small, at least in some species" (p. 145). Dobzhansky concludes the chapter with his development of the idea of microgeographic races, which are, once again, Wright's colonies (nowhere in the text does he call them demes): "The differentiation of the population of a species into numerous small colonies, some of which acquire distinctive morphological characteristics, has received an entirely insufficient amount of attention" (p. 146). He goes on to cite examples of such colonies—really an extension of his review beginning with the land snails of Gulick and Crampton and including data from his own (then unpublished) observations of a European species of ladybird beetle. Here he brings back the continuity-in-diversity argument of the book's previous three chapters. Microgeographic races grade imperceptibly into larger races: "Needless to say, there is probably no sharp dividing line between the microgeographic and the major geographic races; the two merge into each other, and perhaps the former might be regarded as a stage in the development of the latter" (p. 146). And so he concludes, simultaneously paving the way for his discussion on selection:

With the present status of our knowledge, the supposition that the restriction of population size through the formation of numerous semi-isolated colonies is an important evolutionary agent seems to be a fruitful working hypothesis. The action of selection interwoven with and following after that of isolation becomes, as we shall attempt to show in the next chapter, more effective than selection alone is likely to be. (Dobzhansky 1937a, p. 148)

## Dobzhansky on Natural Selection

Dobzhanky's sixth chapter, entitled simply "Selection," starts modestly and is, for the most part, unsurprising (i.e., to readers of the 1980s). In what has long become familiar style and after a brief historical review, Dobzhansky looks at selection in the lab before turning (with customary caution) to the evidence for selection in the wild. Though some aspects of his discussion are interesting, the reader is left totally unprepared for the chapter's conclusion. Upon returning to theoretical population genetics, Dobzhansky throws caution to the wind and abruptly develops a general picture of the evolutionary process that makes far more exciting reading than the material with which he began his discussion of selection.

Dobzhansky clearly links his concept of selection with Darwin's (1859) original formulation. In his definition, the first phrase is the inevitable conclusion that selection must occur given the overproduction of individuals within a species:

If, then, the population is composed of a mixture of hereditary types, some of which are more and others less well adapted to the environment, a greater proportion of the former than of the latter would be expected to survive. In modern language this means that, among the survivors, a greater frequency of carriers of certain genes or chromosome structures would be present than among the ancestors, and consequently the values q and (1 − q) will alter from generation to generation." (Dobzhansky 1937a, p. 149)

The latter statement is the "differential reproductive success" definition of selection that is the current view of selection within the synthesis—and, indeed, of evolutionary biology as a whole—but see Eldredge and Salthe (1984) for a return to a more general definition, where the actual physical (as opposed to simply genetic) deaths of organisms are brought back into the notion of natural selection.

Given the central importance that selection has attained in evolutionary thought within the synthesis, it is of more than passing interest that Dobzhansky states (p. 150): "In its essence, the theory of natural selection is primarily an attempt to give an account of the probable mechanism of the origin of the adaptations of the organisms to their environment, and only secondarily an attempt to explain evolution at large." He then states (p. 150): "Whether the theory of natural selection explains not only adaptation but evolution as well is quite another matter. The answer here would depend in part on the conclusion we may arrive at on the relation between the two phenomena." He mentions (p. 151) the views of R. A. Fisher, "one of the most extreme of the modern selectionists," who, according to Dobzhansky, saw evolution only as "progressive adaptation" forged by selection (in an early version of views later to become commonplace within the synthesis). Fisher's view contrasts strongly in Dobzhansky's mind with a statement he quotes from Robson and Richards (1936) to the effect that, though adaptation via selection cannot be "disregarded as a possible factor in evolution," there is nonetheless little evidence in its favor.

But Dobzhansky does not tell us directly how he feels about the issue. It is already clear that, to him, selection is not the only evolutionary dynamic; drift is important, too. But, for the remainder of the book, virtually all of Dobzhansky's evolutionary theory hinges on the metaphor of Wright's "adaptive peaks," including Dobzhansky's explanation of why there are reproductively isolated units (species) in the natural world. It is thus fair to say that Dobzhansky's integrated picture of the evolutionary process in 1937 saw adaptation—or, more correctly, the problems of climbing, staying astride, and moving between adaptive peaks—as the central problem in evolution. And he does say (p. 150) that the origin of adaptations lies in selection, the only alternative being the inheritance of acquired characteristics. Thus Dobzhansky's position is a complex intermediate between the extreme views of Fisher on the one hand and those of Robson and Richards on the other.

Dobzhansky's review of laboratory and natural evidence for the action of selection considers some of the very same examples that are still widely cited today, including industrial melanism in *Biston betularia* (or *Amphidasis betularia*, as it was then still called). His review of the various "ecological rules" differs a bit from Mayr's treatment in 1942. For example,

Dobzhansky states (p. 165): "It appears that representatives of various families, orders, and even classes inhabiting a given geographical region often undergo there convergent changes, while in another region the same groups seem to vary in a direction which is again similar for all of them but different from the direction observed in the first region." As we shall see in a bit more depth in chapter 3 of the present book, Mayr discusses these rules of geographic variation in the context of within-species variation, explicitly denying that the rules hold at the interspecific level. The difference is hardly crucial here, as Dobzhansky in any case is cautious about claiming too much on behalf of selection from any of his examples.

A brief discussion of the origin of dominance, which contrasts Fisher's views with those of Haldane and Wright (both sets of views involve selection arguments), forms a bridge between what can only be called the conventional (then as now) part of his chapter on selection and the concluding section, which builds up to Dobzhansky's summary vision of the evolutionary process as a whole.

In the section of chapter 6 called "Selection Rate and the Genetic Equilibrium," Dobzhansky returns to mathematical population genetics with this remark (p. 176): "The inadequacy of the experimental foundations of the theory of natural selection must be admitted, I believe, by its followers as well as by its opponents." So he turns to theory, which is "invaluable as a guide for any future experimental attack on the problem" (p. 176). Citing Haldane (1932) to the effect that selection works most quickly on intermediate gene frequencies and rather more slowly if the allele in question is very common or very rare, Dobzhansky writes (p. 177): "If, on the other hand, the initial gene frequency is very small or very large, approaching either zero or one hundred percent of the population, the progress of selection is appallingly slow even with appreciable selection coefficients"—leading to his conclusion (p. 178) that "the number of generations, and consequently the amount of time needed for the change, may, however, be so tremendous that the efficiency of selection alone as an evolutionary agent may be open to doubt, and this even if time on a geological scale is provided." Teggart (1925) has observed that theories of evolution generally see progressive change as inevitable given the mere passage of significant amounts of time; thus, Dobzhansky's remark is relatively unusual, and it is different, as well, from later versions in the synthesis (including Dobzhansky himself in 1951, p. 9, as quoted here in chapter 5), which more closely fit Teggart's dissection of conventional evolutionary theory.

Mutation increases variability, while selection decreases it. In that sense, as we have seen, Dobzhansky thought of the two processes as antagonists. But, of course, selection can go in "the same direction" as "the predominant mutation rate," in which case the two work hand in glove. But if the directions of the two—mutation and selection—are opposed, equilibrium naturally obtains.

When Dobzhansky turns to selection and population size, he moves directly to the work of Sewall Wright. Characterizing Wright's work as "mathematically rather abstruse," he reprints Wright's (1931) figs. 7, 8, 19, and 6 as his figs. 15A–D, respectively, depicting the "distribution of gene frequencies in populations of different size under different selection, mutation, and migration pressures" (Dobzhansky 1937a, fig. 15, p. 181; reprinted here as fig. 2.1). As he explains in simple language the meaning of Wright's mathematics and graphs, Dobzhansky eventually returns (p. 183) to Wright's model of a many-demed species: "The biologically highly significant corollary of the above analysis is that a species, broken up into isolated colonies, may differentiate mainly as a result of the restriction of the population size in these colonies"—a point he, of course, had already made in the preceding chapter as he considered the "scattering of the variability." And here the mere lapse of time does become significant: "Even if the environment is homogeneous for all the colonies, and selection and mutation rates are the same and the initial composition of the populations identical, a sufficient lapse of time will bring about a differentiation" (p. 183). Thus, just before his powerful "conclusions," Dobzhansky returns to the view of species as a series of isolated colonies (though he agrees with Wright that total lack of migration between colonies is probably an unrealistic picture).

Dobzhansky's conclusions start with a recapitulation of his argument so far. Once again we read (p. 185) that "the mutation process, in the wide sense of that term, is adequate to supply the building blocks of evolution." But there are levels, and population-level processes are "transcendental to the mutation process proper." Scattering of the hereditary variability has two effects: it reduces genetic variance within isolated populations, to be sure, but by the same token (as Dobzhansky points out) it increases the variability within a single species that is broken up into such "isolated, self-sufficient" populations. This latter effect is counteracted by migration. Natural selection (p. 186) "combats the excessive accumulation of the gene variants that are unfavorable" and increases the frequencies of rare mutations "that may increase the adaptive level of the species under the static conditions of the milieu." But selection has another role as well:

Natural selection is probably most important when the environment undergoes changes, for it is the sole known mechanism capable of producing a reconstruction of the genetic make-up of the species population from the existing elements. Such a reconstruction may be necessary in order that the species remain attuned to the demands of the environment and escape extinction. (Dobzhansky 1937a, p. 186)

Hence Dobzhansky's dynamics. The problem now is to assess the "relative importance of the different agents." The rest of chapter 6—some five pages—is precisely that: an attempt at synthesis that, not surprisingly, in the second edition is yanked out of the selection chapter and placed in the penultimate chapter ("Patterns of Evolution"), where its climactic sense is perhaps more appropriately situated.

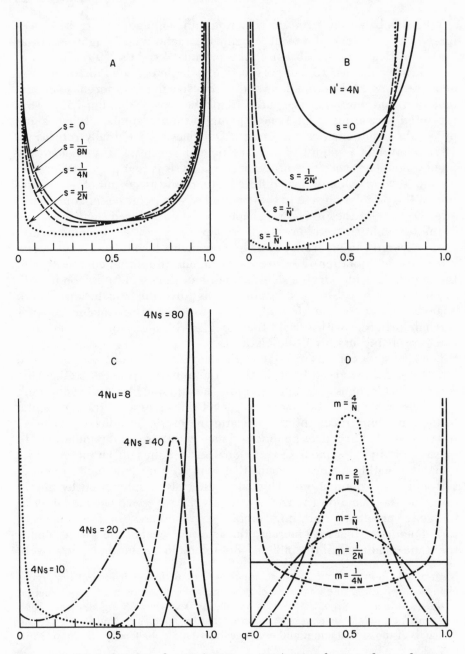

Figure 2.1   Dobzhansky's fig. 15 (1937a, p. 181): Distribution of gene frequencies in populations of different size under different selection, mutation, and migration pressures.

As Dobzhansky puts it (p. 186): "One can hardly eschew trying to sketch some sort of a general picture of evolution. A very interesting attempt in this direction has been made by Wright . . . , whose lead we may partly follow." Follow indeed—the next few pages are so purely Wrightian that, beginning with the next paragraph, the language is lifted directly from Wright (especially 1932) without any particularly close citation to the direct passages that Dobzhansky is recapitulating.

What we have in this section is Dobzhansky's "general picture" of Wright's metaphor of adaptive peaks. His very first paragraph in this discussion (p. 187) is itself a precis of Wright 1932:

In an organism possessing only 1000 genes each capable of producing ten allelomorphs, the number of the possible gene combinations that may be formed is $10^{1000}$. Some, probably a great majority, of these combinations are discordant and have no survival value, but still very numerous ones may be supposed to be harmonious in the different ecological niches of the same environment, as well as in different environments. If the entire ideal field of possible gene combinations is graded with respect to adaptive value, we may find numerous "adaptive peaks" separated by "valleys." The "peaks" are the groups of related gene combinations that make their carriers fit for survival in a given environment; the "valleys" are the more or less unfavorable gene combinations. Each living species or race may be thought of as occupying one of the available peaks in the field of gene combinations. (Dobzhansky, 1937a, p. 187; cf. Wright 1932, pp. 356 ff.)

Still paraphrasing Wright, Dobzhansky goes on to enumerate two "evolutionary possibilities" of the adaptive landscape. First, environmental change may, in effect, modify the field, causing a species to become extinct or to undergo a reconstruction of its genotype, fashioning gene combinations (peaks) once again suitable to the environment. Or, second, a species might "find its way from one of the adaptive peaks to the others in the available field," the field itself imagined as remaining relatively stable. Still following Wright, Dobzhansky sees such shifts as involving "a trial and error mechanism on a grand scale which would enable the species to 'explore' the region around its own peak in order to finally encounter a gradient leading toward the other peak or peaks" (p. 187).

Dobzhansky then introduces two more of Wright's diagrams, one of which is reproduced here as fig. 2.2. Discussing the various possible combinations of evolutionary dynamics given different population structures, all with reference to adaptive peaks, it comes as no real surprise to find as a sixth alternative possibility that a species divided into local races stands the best chance of finding new adaptive peaks: "Wright (1931, 1932) argues very convincingly that the last discussed state of affairs in a species, that is, a differentiation into numerous semi-isolated colonies, is the most favorable one for a progressive evolution" (Dobzhansky 1937a, p. 190).

Dobzhansky then quotes Wright directly to end his discussion:

Wright (1932) formulates his conclusions in the following sentences. "The most general conclusion is that evolution depends on a certain balance among its fac-

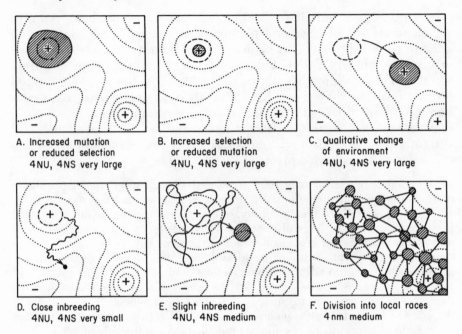

A. Increased mutation
   or reduced selection
   4NU, 4NS very large

B. Increased selection
   or reduced mutation
   4NU, 4NS very large

C. Qualitative change
   of environment
   4NU, 4NS very large

D. Close inbreeding
   4NU, 4NS very small

E. Slight inbreeding
   4NU, 4NS medium

F. Division into local races
   4nm medium

Figure 2.2    Dobzhansky's fig. 17 (1937a, p. 189): Fate of populations of different size and under different conditions in the "field" of gene combinations.

tors. There must be gene mutation, but an excessive rate gives an array of freaks, not evolution; there must be selection, but too severe a process destroys the field of variability, and thus the basis for further advance; prevalence of local inbreeding within a species has extremely important evolutionary consequences, but too close inbreeding leads merely to extinction. A certain amount of crossbreeding is favorable, but not too much. In this dependence on the balance the species is like a living organism. At all levels of organization life depends on the maintenance of a certain balance among its factors." (Dobzhansky 1937a, p. 191, quoting Wright 1932, p. 365)

Wright has, of course, occasionally spoken of selection at various levels other than among organisms within demes. Dobzhansky's espousal of intergroup selection, so different from the hardcore synthesis line of later years, should not perhaps be taken too seriously. But it is yet another example of a more pluralistic earlier phase of thought and so, on the face of it, a view that was able to encompass hierarchical aspects of the evolutionary process. The adaptive landscape became the central metaphor for the synthesis—largely, one might suppose, because Dobzhansky adopted Wright's view of the demic structure of species, thus also the nearly concomitant picture of adaptive peaks so closely associated in Wright (e.g., 1932) with demes and drift. That Dobzhansky's "general picture of evolution" as stated in the closing pages of his chapter on selection (there is some irony in his choice of placement for this passage) is based so utterly on Wright's earlier work is somewhat surprising. But Dobzhansky isn't finished; the adaptive peaks serve a still further and rather crucial role in Dobzhansky's view—one that is purely his own invention. He presents the Wrightian formulations as his general picture, but after a side tracking chapter on polyploidy, Dobzhansky shows us that there is far more to his general picture than he has yet revealed.

Dobzhansky's chapter 7, on polyploidy, is essentially an entertaining potpourri of case histories—including the infamous *Raphanobrassica,* a Russian experimental hybrid between the radish (*Raphanus sativus*) and the cabbage (*Brassica oleracea*) with the risible head of the former and root of the latter. One gets the feeling from this chapter that there has been a break in the action. And though we learn (p. 219) that the prevalence of polyploidy in plants and its rarity in animals "constitutes the greatest known difference between the evolutionary patterns in the two kingdoms," it is difficult to see why the chapter is included at all—difficult to see, that is, until we recall Dobzhansky's near obsession with fairness and completeness. Speciation by "cataclysmic origin" (i.e., via polyploidy; cf. his opening section, p. 192) is the one exception he knows of to his generalization (p. 192) that "the process of species formation is apparently a slow and gradual one, consuming time on at least a quasi-geological scale." Polyploidy is an exception; Dobzhansky duly reviews it, and, since it is rare in animals, he promptly forgets it as the book turns to his general model of species formation, and with it the all-important question of isolation—its nature and origin—as an explanation of the gaps between species whose existence we all acknowledge. And Dobzhansky's explanation of the function of these gaps is quite interesting.

## Dobzhansky on the Origin of Species: The Pros and Cons of Isolating Mechanisms

Dobzhansky's eighth chapter, "Isolating Mechanisms," is perhaps the most purely original body of thought presented in the book. He wrote on the subject in a separately published paper (Dobzhansky 1937b), undoubtedly because he saw his discussion and classification of isolating mechanisms (his term) as new. Dobzhansky's concepts of the function and mode of origin of intrinsic mechanisms that prevent hybrids between groups from reaching "the reproductive stage" provide the key to understanding his views on species in general—what they are, how they came to be, and why they exist at all.

Noting that G. J. Romanes's "oft-quoted maxim" ("without isolation or the prevention of interbreeding, organic evolution is in no case possible") "if taken too literally overshoots the mark" (p. 228), Dobzhansky begins his discussion of isolating mechanisms with a general statement about their importance in evolution. Though he agrees with Moritz Wagner and Romanes that isolation is important in evolution, their disagreement about the reason for this is nearly total. For Dobzhansky, isolation is critical, not as a prerequisite for further change (as Romanes implies), but as a means of consolidation of the status quo which, if anything, acts as a brake on the further accrual of genetic change:

We are confronted with an apparent antinomy. Isolation prevents the breakdown of the existing gene systems, and hence precludes the formation of many worthless gene combinations that are doomed to destruction. Its rôle is therefore positive. But on the other hand, isolation debars the organism from exploring greater and greater portions of the field of gene combinations, and hence decreases the chance of the discovery of new and higher adaptive peaks. (Dobzhansky 1937a, p. 229)

Once again, then, Dobzhansky sees an aspect of the evolutionary process as tensional: isolation has the positive effect of preventing the worthless gene combinations from forming, keeping a species safely ensconced on its adaptive peak. But the price paid for this conservative safety measure is a decrease in genetic variability, hence a limitation on the ability of a species to explore ways and means of attaining new adaptive peaks. Isolation, indeed, conserves in this view.

Late in his career, Dobzhansky wrote (Dobzhansky et al. 1977, p. 168): "The question of why there should be species can thus be answered: because there are many adaptive peaks." In 1937 his position was perhaps a bit less emphatic, but nonetheless similar: species are isolated one from another reproductively, and the reason they are so isolated is to keep them perched atop their respective adaptive peaks. It is, in effect, an argument against mongrelization, as Dobzhansky himself later explicitly acknowledges (see his 1937a, p. 308, and below, this chapter).

Dobzhansky's original classification of isolating mechanisms (1937a, 1941) saw the fundamental dichotomy between those that ensure that no hybrids reach the reproductive stage on one hand and "hybrid sterility" pure and simple on the other. The former category is itself divided into geographic and ecological isolation (the potential mates don't even meet) and what later came to be called "meet but don't (or can't) mate." The classification differs somewhat from Mayr's (1942, pp. 247 ff.; see chapter 3 herein) and his own later treatment (1951), where "don't meet" is contrasted with all the others. Dobzhansky's original classification explains the organization of the remainder of the chapter and the very existence of the next: all isolating mechanisms are covered in the rest of the chapter except hybrid sterility, which receives a chapter (the ninth) of its very own.

Claiming, then, that "the maintenance of species as discrete units demands their isolation" and that "species formation without isolation is impossible" (p. 229), Dobzhansky launches into his review, which, as usual, is fraught with examples from the literature. Geographic isolation—de facto isolation—need not lead to "permanent isolation"; it is, rather, a "temporary" measure and need not lead to a permanent segregation of the groups so isolated (p. 231). The "physiological" isolating mechanisms are the ones of lasting importance, the sine qua non of speciation.

And, in a return to the early argument of his book, Dobzhansky sees the attainment of isolation as a gradational process: races are not as isolated as species, but only incipiently so. Full physiological reproductive isolation—synonymous with the emergence of truly new species (cf. Dobzhansky's chapter 10, p. 312)—is simply the final stage of an essentially continuous process.

Though Dobzhansky's fundamental view sees isolating mechanisms as agents to ensure the purity of species—the mechanism that keeps them on their peaks—and thus serving the role of preventing any two species from hybridizing, he also at times sees the other aspect of such mechanisms essentially as devices to ensure the mating between conspecific individuals. The distinction between the two views on the significance of such mechanisms is rather critical, as we shall see in later chapters. Following the analyses of H. E. H. Paterson (1978, 1980, 1981, 1982), the positive recognition of conspecific mates is not at all the same thing as a mechanism that acts solely to prevent interspecific hybridization. For the moment, it is only important to note that Dobzhansky's clearest statement of the importance of specific mate recognition systems (Paterson's term; cf. Paterson 1978) comes under his section on sexual isolation:

Sexual recognition marks of various kinds (specific scents, colorings, sounds, and various behavior patterns grouped under the name courtship) enable the individuals of either sex to discern potential mates. Any incongruity between the mating reactions of two groups of individuals may engender sexual isolation. (Dobzhansky 1937a, p. 235)

But there is no doubt about where Dobzhansky's real interest lies. To Dobzhansky, the fact that species differ in their sexual behavior is interesting primarily because it is a barrier (admittedly imperfect) to hybridization between two species, a means primarily of preventing the formation of disharmonious gene combinations.

On page 254, Dobzhansky starts to develop his "working hypothesis" on the origin of isolation. First he says (p. 255) that single mutations simply won't suffice. The problem is not only isolation of group B from group A but also the ability of organisms *within* the new group B to breed among themselves. In any case, he argues (p. 256), isolating mechanisms involve whole systems of complementary genes. He then goes on to build what is still usually construed to be the conventional view of the synthesis (cf. Altokhov 1982, p. 1168: "Reproductive isolation, an important criterion of species, is viewed in this model as only a by-product of such differentiation"). Physiological isolating mechanisms are conventionally seen as consequences, side effects (or, simply, "effects" in the language of Williams 1966 and Vrba 1980). The mere accrual of genetic differences between geographically isolated populations in species leads (or may lead) to the development of these intrinsic physiological isolating mechanisms. That allopatric speciation had yet to be established as the virtual single model of speciation acceptable in the synthesis is seen in Dobzhansky's statement (p. 257) that "The assumption that geographical isolation is a *conditio sine qua non* of species formation is, nevertheless, not a necessary one"—ecological isolation remaining a distinct alternative possibility to Dobzhansky. But geography is nonetheless central to his argument:

A geographical isolation of parts of a population may be followed by the appearance in the subgroups of inheritable changes that engender a permanent isolation between them. It follows that we may witness in nature isolating mechanisms *in statu nascendi*, when some individuals are already isolated and others not yet isolated from other species. (Dobzhansky 1937a, p. 257)

Thus, Dobzhansky's model of the origin of isolating mechanisms stresses the continuity and gradualness of genetic divergence between geographically isolated populations.

But Dobzhansky is aware that he has created a problem. He began by strongly asserting the nature and function of isolating mechanisms: to keep species apart, to keep them on their adaptive peaks, and to prevent the formation of Wright's unharmonious constellations of genes. That is their evolutionary role—the benefit they provide to species—albeit at the cost of some evolutionary plasticity. Yet the theory Dobzhansky adduces to explain their origin sees isolating mechanisms as accidental byproducts of geographic isolation, mere accumulation of sufficient genetic divergence (whether by selection or scattering of the variability, i.e., genetic drift) which accrues to the point where reproductive isolation results. What vexes Dobzhansky is that his general theory of evolution up to this point sees a neat interplay between statics and dynamics: each aspect of the process that has a specifiable role—or positive benefit— should have some underlying cause other than mere chance (or at least chance alone). After all, if it is beneficial for a species to be isolated, there should be some mechanism more definite and deterministic than mere accident underlying the origin of its isolation. Dobzhansky here would dearly love to invoke selection as the root of isolation:

The spread within a population of genes that may eventually induce isolation between populations is probably due to their properties other than those concerned with isolation. What these properties are is a moot question, and here is the weakest point of the whole theory. . . . Isolation is in general a concomitant of the genetic differentiation of separate populations. It may be noted, however, that only those genetically distinct types that have developed isolation can subsequently coexist in the same region without a breakdown of the differences between them due to crossing. Therefore, isolation becomes advantageous for species whose distributions overlap, provided that each species represents a more harmonious genetic system than the hybrids between them. Under these conditions the genes that produce or strengthen isolation become advantageous on that ground alone, and may be favored by natural selection. This may at least be a partial solution of the difficulty stressed above. (Dobzhansky 1937a, p. 258)

This paragraph nicely summarizes Dobzhansky's dilemma and "partial solution," which has since come to be called "reinforcement": partial reproductive isolation attained in allopatry (so later versions of the argument go) may well be strengthened (i.e., reinforced) when sympatry is regained and selection acts against the hybrids formed by the partially reproductively isolated groups. Dobzhansky does not admit *why* this problem is the "weakest point of the whole theory." But there can be no answer other than the obvious: isolation is adaptively advantageous. It keeps species on their peaks. Therefore isolation ought to be achieved under the aegis of selection. That this problem was acute in Dobzhansky's mind becomes all the more apparent in the second edition, where (as we shall shortly see) he greatly strengthens his selectionist argument on the origin of isolating mechanisms.

Chapter 9, devoted entirely to hybrid sterility, completes Dobzhansky's survey of isolating mechanisms. Like the earlier chapter on polyploidy, this chapter seems something of an aside, a bit of a departure from the main argument included for the sake of completeness (as opposed to the chapter on polyploidy, which is there to take into account the one argument against continuity and gradual diversification along the path of speciation). We do have yet another claim for continuity, in reference to "races A and B" of *Drosophila pseudoobscura,* which, at the end of the second edition (1941, p. 347), Dobzhansky says have all the earmarks of distinct species: "The gap between the interracial differences causing the sterility of hybrids and the intraracial variations is thus bridged" (p. 286). Otherwise, chapter 9 contributes little to Dobzhansky's evolutionary theory per se.

### Species as Natural Units

Dobzhansky concludes his book with a tenth chapter entitled "Species as Natural Units." It is a chapter that essentially consolidates his previous narrative, while at the same time by no means serving as a summary of the book. (In fact, the chapter, and thus the book, ends rather disappointingly on a brief discussion of asexual species, once again reflecting Dobzhansky's desire to be complete in his consideration of nature.) The chapter adds but a single element to the overall argument—but it is a powerful addition, one that utilizes the landscape imagery to explain the entire evolutionary history of life. It is in this chapter that Dobzhansky tells us how that "sign of equality" can be placed between microevolution and macroevolution.

Dobzhansky begins his final chapter on an unexpected (and most welcome) note, giving us his views on systematics. His reason for so doing soon becomes clear. Just as Darwin was able to argue (1859, chapters 4 and 13) that the natural system described by taxonomists only made sense in light of "descent with modification,"[4] Dobzhansky makes much the same point. He begins by reminding us of something he said before: there are two aspects to evolution—diversity (which has a continuous base) and discontinuity—and discontinuity, being less obvious, has been somewhat overlooked. Discontinuity is an ontological aspect of the living world; it is a strictly empirical observation. Moreover, Dobzhansky asserts, even though Darwin supplied the hypothesis of common descent as an explanation of the "natural classification" of organisms, the plain fact is that the construction of phylogenetic trees and the findings of paleontology have essentially effected no change in the efforts of systematists to classify the natural system: "And yet the classification has continued to be based chiefly on morphological studies of the existing organisms rather than of the phylogenetic series of fossils" (p. 304). We need only to make the assumption that "similarity between the organisms is a function of their descent" (p. 305), though he admits that this procedure leads to difficulties. But the difficulties are "more abstract than real": the system existed *before* Darwin (indeed, it exists in the real world independent of our ideas of it), and the major source of inconstancy of classification (i.e., according to Dobzhansky; today we would call it instability) turns out to be merely changes in the assignment of rank to taxa and not so much in the recognition of taxa per se.[5]

But there is, Dobzhansky says, "a single systematic category which, in contrast to others, has withstood all the changes in the nomenclature with an amazing tenacity" (p. 306). This, of course, is the species—a nonarbitrary affair simply because "no category is arbitrary so long as its limits are made to coincide with those of the discontinuously varying arrays of living forms" (p. 307). In addition, species, unlike other "categories," have unique properties.

This much established, Dobzhansky turns to the "Genetic Basis of Classification" and the last really significant discussion in the book insofar as his overall argument is concerned. First he asks us (p. 307) to imagine a world in which "all possible gene combinations are represented by equal numbers of individuals. Under such conditions no discrete groups of forms and no hierarchy of groups could occur, since the single gene differences producing striking phenotypical effects, like some of the mutations in *Drosophila*, would be the sole remaining source of discontinuity. Disregarding these, the variability would become a perfect continuum."

And, of course, as he has already made abundantly clear, the real world is by no means like this fantasy. In the several paragraphs spanning pages 307 and 308, Dobzhansky in effect summarizes the early synthesis. (I urge the reader to consider these paragraphs in full in the original.) He tells us that only a fraction of the possible gene combinations "is realized among the living individuals." And these gene combinations "are grouped together into more or less compact arrays"—and clustered, moreover, onto adaptive peaks.

The discontinuity and the hierarchical character of the empirically observed variation may be viewed as a corollary of the particulate structure of the hereditary materials. . . . Each race, species, genus, or any other group embraces a certain array of gene combinations attached to an "adaptive peak," or to several neighboring peaks. . . .

The conclusion that is inexorably forced on us is that the discontinuous variation encountered in nature, except that based on single gene differences, is maintained by means of preventing the random interbreeding of the representatives of the now discrete groups. This conclusion is evidently applicable to discrete groups of any rank whatever, beginning with minor races of a species and up to and including classes and phyla. The development of isolating mechanisms is therefore the *conditio sine qua non* for emergence of discrete groups of forms in evolution. (Dobzhansky 1937a, p. 308)

Here is Dobzhansky's extension of the imagery of the adaptive landscape to taxa *above* the species level.[6] While the lower-level genetic statics (e.g., mutation) are not explicitly incorporated here—nor are the intricate interplays between population size, selection, mutation, and drift— the larger-scale imagery of species perched on peaks, with isolation to prevent the promiscuous formation of freaks, shines through, and by explicit extension, higher-ranked taxa, also discrete entities, fall into the fold. In other words, Dobzhansky is saying that those processes he has invoked to produce the origin of new, isolated species are the very mechanisms that produce collections of closely related species—the "categories" of higher rank. Much else belongs to the synthesis—even in his 1937 edition but also in his later writings and especially those of Simpson and Mayr—but (in conjunction with his earlier discussions of lower-level phenomena) Dobzhansky's passage (pp. 307–8) is as effective a summary of the early synthesis as can be imagined. Nor should it be overlooked that the fundamental view of nature to which Dobzhansky adheres relies to a very considerable extent on a single prior publication: Wright's brief paper of 1932.

There is one additional point to be made about this key passage: discontinuous variation in the organic world, Dobzhansky says, is a consequence of the genetic make-up of organisms. Particulate inheritance (including mutation) itself does lead to discontinuity, but in the population context it quickly dissolves to utter continuity. Thus, a higher-level barrier to continuity—the isolating mechanisms—is required, else we get the "mass of freaks." Just why the world should not be one continuous interbreeding pool (with selection simply weeding out those inharmonious combinations from each generation) Dobzhansky doesn't say. It is not a particularly easy problem. But, once again, species as isolated reproductive units have a real purpose in Dobzhansky's vision of the evolutionary history of life.

Dobzhansky concludes his first edition with a discussion of sexual, then asexual, species. Though discontinuity preoccupies much of the book, Dobzhansky reminds us that, in the light of evolutionary theories (p. 309): "such concepts as race, species, genus, family, etc., have come to be understood as nothing more than degrees of separation in the process of a gradual phylogenetic divergence." Yet there is something special about species.

Then Dobzhansky discusses the difficulty in defining species, coming up in due course with his own definition (partly cited above and originally proposed in a separate paper, Dobzhansky 1935). To Dobzhansky, a "species is that stage of the evolutionary process at which the once actually or potentially interbreeding array of forms becomes segregated in two or more separate arrays which are physiologically incapable of interbreeding" (1937a, p. 312). This definition, he admits, deprives the systematist of a "fixed yardstick" with which a given species might be defined and recognized. As I have already noted it really is almost as if Dobzhansky, looking up from the level of the gene, sees species as discrete entities, real and concrete. But viewing species per se, they become a stage on a continuum and one that can be defined by what can only be called a threshold (p. 315): "We may conclude that if the species separation is defined as a stage of the evolutionary process at which physiological isolating mechanisms become developed, the species so defined and the species of systematists will largely coincide." Dobzhansky's definition sees species as a sort of class, but his pragmatic view of nature sees them as real and tolerably discrete entities.

Finally, asexual species pose a paradox. There are still "biotypes" among them; they "obviously do not embody all the potentially possible combinations of genes" (p. 320). Their very existence troubles Dobzhansky—because his entire theory hinges on the importance of isolation to prevent total gene mixing and to keep sexually reproducing species on their adaptive peaks. Asexual organisms should have no need of such a mechanism. Yet they are (p. 320) "clustered around some of the 'adaptive peaks' in the field of gene combinations, while the 'adaptive valleys' remain more or less uninhabited." Always honest, always striving for completeness, Dobzhansky ends his book on a bit of a sour note—one that was not removed until judicial editorial rearrangement ended the third edition (1951) on a more general theme.

## THE SECOND EDITION: *GENETICS AND THE ORIGIN OF SPECIES* IN 1941

Dobzhansky's second edition is, of course, substantially the same book as the original. The text is little altered in most places, and many of the original illustrations and tables remain. His fifty-nine additional pages consist mostly of updated examples drawn from the literature of the intervening four years. Yet there is a pattern of deletion, addition, reorganization, and modification that subtly but definitely strengthens several of the elements of his original picture of the evolutionary process.

The most important change lies in the role Dobzhansky sees for selection in evolution. He deletes some of his earlier skepticism about the all-powerful nature of selection (and the concomitant importance of drift). For example, in his discussion of the "scattering of the variability" in the first edition, Dobzhansky had said (1937a, p. 131): "In fact, as shown below, these events may take place even in spite of natural selection; genes which produce slight adverse effects may reach fixation and more advantageous genes may be lost." This statement is simply deleted from the end of the same paragraph in the second edition (1941, p. 164), which otherwise remains unchanged. A few pages later (1937a, p. 133; 1941, p. 168), where he first introduces Wright's vision of "species whose population is subdivided into numerous colonies of different size," he had originally said (1937a, p. 134): "The conclusion arrived at is an important one: the differentiation of a species into local or other races may take place without the action of natural selection." This entire section is now utterly gone in the second edition—not because Dobzhansky has dropped Wright's imagery, but because he has rearranged his argument, deleting the multideme picture of species from the two chapters on population dynamics (i.e., the chapters on within-population variation and on selection) and reserving all this material for his one new chapter, "Patterns of Evolution," chapter 10 in the second edition. But the comment denigrating the importance of selection does not reappear in that new chapter; it has simply been quietly dropped.

The net effect of Dobzhansky's rearrangement is that Wright seems to have a less detectable influence in the second edition, even though the only real change is a shift away from seeing drift part and parcel as nearly equally important as selection in effecting changes in gene frequencies within populations. Instead of the strong pro-drift attitude of the remark just quoted from the first edition, Dobzhansky tones his statements way down, merely saying (1941, p. 168): "If population sizes in most species tend to be small on the average, the scattering of the variability and the random variations of the gene frequencies will loom important as evolutionary agents." He then reviews the ambiguous evidence for population size in nature, concluding as he had done earlier that many species are indeed divided up into numerous colonies. Drift remains important, but the paragraphs concluding this chapter in the two editions differ drastically, as Dobzhansky's enthusiasm for what he now (1941) calls drift has been sharply curbed.

The "winner" here, then, is selection. Yet Dobzhansky's chapter on selection is actually shorter than it was in the original edition. He has excised his final discussion (where he gives his general picture of evolution in the first edition) and placed it in the new chapter 10, understandably reserving his conclusions for a point closer to the actual end of the book. The selection chapter is rearranged and updated somewhat but otherwise little changed. It isn't until we reach chapter 8—"Isolating Mechanisms"—that Dobzhansky explicitly strengthens his statements on the importance of selection in evolution.

As we have seen, Dobzhansky's view of the role of isolating mechanisms as a preventive against the formation of unharmonious gene combinations conflicted with his conclusion that physiological isolating mechanisms seem to arise as a mere byproduct of geographic differentiation. Thus, though he suggests reinforcement as a partial solution to his dilemma, he openly admits in the first edition that the disparity between the evolutionary role and the mode of origin of physiological isolating mechanisms is the weakest part of his theory. In his second edition, Dobzhansky doggedly confronts the problem head on. Once more reviewing the literature, he finds (1941, p. 283) that "the thesis that the development of physiological isolating mechanisms is preceded by geographical isolation finds a good deal of evidence in its support in observations on the organic variation in nature." Yet (p. 284) "in the opinion of the writer, geographical separation in itself does not guarantee the eventual advent of a reproductive one (cf. Goldschmidt 1940)." He asks (1941, p. 284), "What causes engender the development of physiological isolation between natural groups?"

Dobzhansky then elaborates a model that sees isolating mechanisms as "*ad hoc* contrivances which prevent the exchange of genes between nascent species, rather than [as] incongruities originating in accidental changes in the gene functions." Citing his own 1940 paper, Dobzhansky elaborates a model that sees the development of physiological isolating mechanisms thoroughly as "a product of natural selection." The model (developed in Dobzhansky 1941, pp. 285–86) hypothesizes mutant forms in either or both of two incipient species—mutations that render their carriers less likely to mate with representatives of the other species. Selection favors the spread of the mutant forms (which mate mostly or wholly with conspecifics) over the nonmutant forms, which are further imagined to show a somewhat reduced viability of hybrid offspring. From such a scenario, it is easy to imagine selection against further hybridization: "Once an incipient physiological isolation has become initiated, natural selection will tend to strengthen it and eventually make the isolation complete" (1941, p. 286).

But, Dobzhansky candidly admits, "it is . . . not clear how physiologically isolated species could develop in noncontiguous territories, such as oceanic islands, where opportunities for hybridization are not available" (p. 287). But he has gone a long way toward solving his problem—isolating mechanisms are "*ad hoc* contrivances"—and now Dobzhansky has at least partly answered his own question about how we can imagine selection developing these useful preventives against interspecies hybridization. Thus, Dobzhansky strives to solve the one inconsistency he himself acknowledged in the first edition, and he does it by strengthening and expanding the role he sees for selection in the evolutionary process.

Also strengthened over the first edition is Dobzhansky's continuity argument where mutations (in the broad sense, embracing chromosomal changes) are seen as the ultimate source of diversity in nature. Gone are some of his more cautious views of the first edition, and added are more encompassing and assertive claims, the latter frequently in conjunction with a determined response to the views of Richard Goldschmidt, whose book, *The Material Basis of Evolution*, had appeared the year before.

(Indeed, on pp. 336 ff. of the second edition, Dobzhansky openly admits that Goldschmidt is egging him on. Repeating his statement of the first edition that "a geneticist can hardly eschew trying to sketch some sort of a general picture of evolution," he adds, "the more so since Goldschmidt . . . published his challenging theory of systemic mutations.")

Dobzhansky also updates his discussion of species definitions, noting (1941, p. 374) with unmistakable pleasure that a number of biologists (including Ernst Mayr) had adopted his species definition "or variants which retain the basic idea." Species, of course, remain for Dobzhansky real entities in nature: "Species so defined are tangible natural phenomena" (p. 374).

But it is in the single new chapter of the second edition (chapter 10, "Patterns of Evolution") that we receive a coherent update and summary of Dobzhansky's evolutionary theory—plus intriguing discussions of the data of systematics and paleontology, tellingly interpreted in the context of his theory. He also adds comments on "specialization and rudimentation" and several other topics (including evolutionary patterns in humans), none of which particularly bears on an exegesis of the essentials of the synthesis. But it is in the first thirteen pages of the chapter that, once again, Dobzhansky presents his own synthesis.

And that's the way Dobzhansky himself puts it—synthesis—as he begins his discussion. He lists (1941, p. 331) "gene mutation, chromosome changes, restriction of the population size, natural selection, and development of isolating mechanisms" as "the known common denominators of many, if not all, evolutionary histories." [Simpson's list (1944, p. 30; see chapter 3 of this book) of the "determinants" of evolution is similar but not identical: "variability, rate of mutation, character of mutations, length of generations, size of populations, and natural selection." Note the absence of drift on both lists, though it is certainly implied in both in "population size".] How they all work together forms the subject matter of Dobzhansky's new chapter.

He then merely repeats the section "Selection in Populations of Different Sizes "(rewriting the initial paragraph) which was included at the end of the selection chapter in the first edition. Also included are Wright's graphs (see fig. 2.1 above; this section, as we have already seen, is a brief summary of Wright's analysis of the "distribution of gene frequencies in populations of different size under different selection, mutation and migration pressures," Dobzhansky 1937a, p.181).

Next, what was labeled "Conclusions" in the first edition and presented as Dobzhansky's general picture of evolution (also in the selection chapter of the first edition) appears now as "Progressive Evolution" (1941, p. 336). First commenting (p. 336) that "inferences regarding macroevolution which may be drawn from genetic data are of necessity extrapolations," he repeats in somewhat altered form his precis of Wright's concept of the adaptive landscape. In an important addition to his earlier commentary, Dobzhansky touches on the difficulties inherent in transitions from one adaptive peak to another. Simpson (1944; see below) dwells in detail on the problem of how valleys are crossed and new peaks climbed. Dobzhansky addresses it merely by going back to his original text, with a summary of Wright's diagrams and a survey of the various possibilities, once more concluding that the sixth possibility, which emphasizes the "division (of a species) into local races," is the most favorable for progressive evolution (1941, p. 340).

But Dobzhansky cuts short his exposition of this model (which ended his discussion of selection in the first chapter) and instead, in quick succession, addresses first systematists and then paleontologists. On page 342, he remarks: "Systematists are inclined to regard the geographical differentiation of a species into local races, rather than a high intrapopulational variability, as the precursor of the formation of new species"—obviously a point that fits in well with his (and Wright's) notions. He then calls for a survey of the "available data on their [i.e. animals' and plants'] ecology, breeding behavior, and especially on the effective population sizes"—something of a harbinger of Mayr's book (1942) which appeared shortly thereafter.

But it is Dobzhansky's application of his theory to the fossil record that is particularly arresting. He concludes his section on "progressive evolution," as I shall complete my analysis of the first two editions of *Genetics and the Origin of Species*, with two paragraphs that recognize the existence of "gaps" between groups in the fossil record and attempt to account for them as a natural consequence of evolution and not as a mere function of the vagaries of fossilization. Much of what he says here so succinctly is to be found in Simpson's notion of "quantum evolution," including the idea that the species involved in the transitions from peak to peak will probably be rare (cutting down their chances of preservation and discovery as fossils), with relatively smaller population sizes, and displaying relatively faster rates of evolution. Included with these two paragraphs is a footnote contrasting (accurately, I believe) the superficially similar ideas of Goldschmidt (1940) and the paleontologist (and saltationist) Schindewolf (1936), whose names are somewhat linked in the following two paragraphs:

Certain generalizations arrived at by palaeontologists should be seriously pondered by geneticists. The fossil record indicates that the tempo of evolution within a phyletic line is not uniform in time: periods of an explosive proliferation of new forms are succeeded by a more gradual development. A circumstance that is most irksome to a student of phylogenetic histories is that major evolutionary advances, the first appearances of new families, orders, and classes, seem to occur suddenly, with few or no intermediates between the new groups and their putative ancestors being preserved as fossils. The "missing links" are, indeed, seldom found. Schindewolf . . . in his brilliant and provocative discussion of the bearing of palaeontological findings on theories of evolution, stresses this sudden emergence of the radically new organic forms and infers that evolution takes place in part by what seems to correspond to Goldschmidt's systematic [sic] mutations. The facts at hand can, however, be accounted for without recourse to Goldschmidt's assumptions.

Transition from one adaptive peak to another quite remote from the first in the field of gene combinations entails a radical reconstruction of the genotype. In general, the more radical the reconstruction the less likely are the intermediate stages of the rebuilding process to represent harmonious genotypic systems. Goldschmidt and Schindewolf have correctly pointed out (as others have done before them) that when observing two sufficiently different types of organization, each of which is harmonious in itself, it is frequently difficult to visualize functional systems intermediate between the two. As stated above, the roads leading from one adaptive peak to others usually cross adaptive valleys. The "missing links" are, therefore, expected to be creatures able to subsist only in certain special environments, and hence rare in nature. Rare species are evidently seldom preserved and discovered as fossils. In as much as the effective population sizes in species that are rare or confined to strictly circumscribed types of habitat are likely to be smaller than in common and widespread species, the rates of evolution may be greater on the average in the former than in the latter. In a sense, evolution may at times defeat its own ends: successful types multiply, spread, conquer vast territories, but in so doing they run the risk of losing the very population structure which enabled them to become successful. (Dobzhansky 1941, pp. 343–44)

Note once again Dobzhansky's attraction to tensional conflict—the pros and cons of following a particular evolutionary "strategy." Here, one may win the game at the cost of losing the very characteristics that conferred success in the first place.

In turning now to the books of Mayr (1942) and Simpson (1944), we by no means abandon these two editions of *Genetics and the Origin of Species.* For as little as Mayr's and Simpson's works overlap in theme and content, both share a great deal of each with Dobzhansky's great work.

## NOTES

1. Parts of my critique of Mayr (1942) appeared as the introduction (Eldredge 1982b) to a newly reissued edition of that work. The first edition of *Genetics and the Origin of Species* (Dobzhansky 1937a) was reprinted in 1982, with an introductory essay, synopsis, and critique by Gould (1982a). Simpson's (1944) *Tempo and Mode in Evolution* reappeared in 1984. Naturally, my exegesis of these books, essential to my own narrative, cannot replace an actual firsthand reading of the texts themselves; each continues to repay careful study. In the 1980s, we know these biologists primarily through their later works, and the excitement and freshness of their vision in the late 1930s and early 1940s—their essential creativity—are necessarily dampened to some extent in those later works.

2. Ernst Mayr (personal communication) corroborates this story and has said that he helped persuade Dobzhansky not to call his book simply the fourth edition of *Genetics and the Origin of Species.*

3. The distinction between taxonomic categories (which are classes, such as species, genera, orders, phyla, etc.) and taxa (which are particular instances or entities belonging to such a class—individuals in the sense used later in this book—such as *Homo sapiens*, Rodentia, Arthropoda) was not to be made explicit until Simpson (1963). The problem reappears in the present chapter (see note 5 below) and is discussed again in chapter 3.

4. Indeed, the complexly nested hierarchical array of similarities and the parallel array of taxa (the hierarchies of homology and taxa, respectively, discussed in chapter 7 of this book) have long been used as evidence in support of the very notion of evolution. In the context of the "hypothetico-deductive" approach to science, it is equally valid and perhaps more

useful to see these twin hierarchies (presented in this book as part of the historical results of the evolutionary process—see Eldredge and Salthe 1984; Vrba and Eldredge 1984) as the predicted pattern to be found in nature *if* evolution is "true." This position is stated in more complete, albeit informal, terms in Eldredge 1982c, chapter 2.

5. As I have already noted (note 2 above), the distinction between category and taxon had yet to be made when Dobzhansky was writing. The word he uses here is actually *group*. The issue is important, because part of the basis for criticism of the synthesis in recent years is that it has treated some biological entities as if they were classes rather than spatiotemporally bounded entities (i.e., individuals). The criticism is not based, however, on the confusion between categories and taxa—which, however important, is surely not a fatal flaw in early writings of the synthesis. Whenever *category* is encountered, the reader may safely assume the writer means *taxon*.

6. Eldredge and Cracraft (1980, p. 251), in a brief treatment of how Wright's within-species imagery became transposed to embrace taxa of higher rank, noted that Mayr (1942) had used the phrase in its extended sense, but only in passing. Simpson's *Tempo and Mode* (1944) seemed the earliest full discussion of the extended version of the metaphor of the adaptive landscape. But these last two paragraphs from Dobzhansky (1937a) make it abundantly clear that the expansion had already been accomplished in his first edition, though by no means as polished or pervasive in its claims as the version appearing in his third edition (1951, p. 9, quoted in chapter 5 herein).

# 3

# Systematics, Paleontology, and the Modern Synthesis

"The systematist who studies the factors of evolution wants to find out how species originate, how they are related, and what this relationship means. He studies species not only as they are, but also their origin and changes. He tries to find his answers by observing the variability of natural populations under different external conditions and he attempts to find out which factors promote and which inhibit evolution. He is helped in this endeavor by his knowledge of the habits and the ecology of the studied species." (Mayr 1942, p. 11)

Ernst Mayr, a systematist and founding father of the synthetic theory, has recently (Mayr 1980b) assessed the role played by the field of systematics in general in the emergence of the synthesis. Mayr (1980a, 1980b; 1982, chapter 12) actively opposes the conventional supposition that the synthesis is the product of three phases of development: (1) resolution of early difficulties raised in the early days of genetics, largely through the work of Fisher, Haldane, and Wright; (2) the publication of Dobzhansky's *Genetics and the Origin of Species* (1937a), which fused concepts of the genetics of populations with the mainstream of Darwinian thought; and (3) the demonstration by systematists (e.g., Mayr 1942), paleontologists (e.g., Simpson 1944), and practitioners of various other biological disciplines that the data of their respective fields are consistent with genetic principles. (See, for example, Shapere 1980, p. 398, for such a view of the historical development of the synthesis.) It is Mayr's view (e.g., 1980a, 1980b) that these various nongenetics disciplines played a more vital, vigorous and active role than such "me-too-ism" implied by phase 3 above in the conventional view. There is, no doubt, something to be said for this claim, though the historical question per se is not germane to the present inquiry. But Mayr's (1980b, pp. 127 ff) list of the contributions he feels systematists made directly to the new synthesis is relevant as it suggests a guide to our understanding of Mayr's own important contribution—*Systematics and the Origin of Species* (1942, reprinted in 1982).

Mayr lists the following contributions of systematics to the emerging synthesis: (1) "population thinking," (2) "the immense variability of populations," (3) "the gradualness of evolution," (4) "the genetic nature of gradual evolution," (5) "geographic speciation," (6) "the adaptive nature of observed variation," (7) "belief in the importance of natural selection," and (8) the notion (shared with paleontologists) that "macroevolutionary phenomena" are interpretable in terms of "gradual evolution" (i.e., as opposed to saltational models—Mayr 1980b, p. 134). All these topics, and more, are well developed in the pages of Mayr's *Systematics and the Origin of Species*.

Just as Dobzhansky (1937a) claimed not to be discussing the phenotype and its role in evolution per se, Mayr (1942, p. 70) wrote in reference to Dobzhansky's 1941 edition: "In order to avoid duplication, I have attempted to reduce to a minimum, in my own presentation, all references to genetic material." Whereas Dobzhansky (1937a, 1941) started with the "physiology of the gene" and worked his way up through populations into species and (to a lesser extent) among-species phenomena, Mayr's chief concern throughout his book is with the internal anatomy of species, starting from the vantage point of the "characters" of taxonomy—in other words, from the standpoint of the phenotype. It is the phenotype, after all, through which selection acts to modify the genetic composition of populations.

Though the overlap between Mayr's and Dobzhansky's books is great— both, for example, see diversity (genetic and phenotypic) and disconti- nuity as the twin, coequal grand themes in evolution—nonetheless Mayr's perspective as a systematist creates a different feel, provides an utterly different array of information about the biological world, and, most inter- estingly, leads him to some different conclusions about both the nature of some evolutionary units and the processes that produce them.

## SYSTEMATICS AND THE ORIGIN OF SPECIES

In a sense, *Systematics and the Origin of Species* represents two books— or at least two separable themes to be developed by Mayr later in his career as separate books. The first theme is an inquiry into the nature of species, how they evolve, and how their evolution is related to population genetics on the one hand and larger patterns of evolution on the other. This, of course, is the prime focus of the book and the theme for which the book is remembered. Mayr's (1963) *Animal Species and Evolution* (the abridged version of which appeared in 1970 as *Populations, Species, and Evolution*) is a later, expanded and modified version of this theme.

The second theme, muted and occupying far less space in the 1942 work, is also an inquiry into the nature of species, and focuses on how they might be recognized, how their relationships with other species might be analyzed, and how the patterns emerging from such an analysis might be classified. It is, in short, a brief overview of the practice of systematics itself, including an evaluation of the state of that science in the early 1940s and a brief statement of the principles of systematics. Mayr later wrote an entire volume on the subject with Linsley and Usinger (1953), returning to it once again in his (1969) *Principles of Systematic Zoology*. Mayr is as well known for his espousal of the "new systematics" (and his views on numerical taxonomy and cladistics) as he is for his views on speciation and its relation to the entire evolutionary process. His views on systematics have also changed over the years. The connection between his ontological concerns with the evolutionary process and his epistemological concerns with systematics resides in his concept of the very nature of species, which dictates both how we think they evolve and how we should proceed in the business of recognizing them. This second, subsidiary theme of systematics is an important aspect of the book but takes us too far afield from the synthesis per se to follow here. I have commented at length on Mayr's views on systematics in 1942 in my introduction (Eldredge 1982a) to the reprinted edition of Mayr's book.

## Mayr on Species

The nature of species forms the crux of the book, the point of departure for the development of Mayr's ideas on both systematics and evolutionary theory. In this context it is perhaps surprising that species remained undefined throughout the first hundred pages. The word *species* occurs throughout, and a very good idea of what Mayr is driving at emerges at the outset, in chapter 1, where the old "typological" species concept is vigorously assailed, to be replaced by the notion that species are inherently, internally variable. The succeeding three chapters document the nature and extent of this variation, and, finally, in chapters 5 and 6, Mayr summarizes the argument on variation and adds a second ingredient, reproductive integrity, the source of cohesion that at once unites the populations comprising a species and isolates that species from other groups. Chapter 7 begins Mayr's exposition of how new species arise and includes his characterization of allopatric speciation. Chapter 8 considers (and essentially rejects) alternatives to the notion of geographic speciation, and chapter 9 looks in detail at the "biology of speciation," those factors promoting and inhibiting speciation.

Thus, the ontological questions of the evolutionary process that Mayr addresses are divided into *what* (what are species?—chapter 1, in part, and chapters 2 through 6) and *how* (how do species originate?—chapters 7 through 9). This organization is not merely a logical choice devised for purely didactic purposes. For Mayr, what species are largely explains how they come into being.

To Mayr, there are two crucial issues involved with the very definition of species, and therefore with the explanation of how they evolve: morphological diversity and discontinuity, first explicitly discussed on page 23 and reiterated throughout the book. Recall that these two themes are the very same central issues that Dobzhansky (1937a, 1941) saw for evolution in general. To Mayr, there is variation within and between species, and theories of speciation must embrace such variation, explaining both its nature and its origin. But there is also the problem of discontinuities between species—the "bridgeless gaps," a phrase much used in the second half of the book and taken with approbation (for the most part, but see Mayr's p. 114) from the writings of Richard Goldschmidt on the gypsy moth *Lymantria*. A theory of the nature of species and their mode of origin must also address the problem of the origin of these discontinuities, these bridgeless gaps. Mayr is at times rather vague about what these gaps really are: sometimes he clearly means reproductive isolation, but at others he uses the phrase strictly in allusion to the nonintergrading patterns of morphological (and presumably genetic) differences between species—the sort of differences seen by fellow systematists and also what Goldschmidt meant by the expression "bridgeless gaps."

Mayr is at pains to show that phenotypic differentiation within species underlies phenotypic differences among species as a smooth continuum. But the very gist of this outlook of smoothly gradational continua collides with the notion of discontinuity, which itself is not discussed in detail until the fifth chapter. The dilemma is real, for without these discontinuities species do not exist for Mayr in any meaningful sense of the word, and there would simply be nothing there to explain. His treatment of the entire problem implies a recognition that if one goes too far in embracing the principles of natural selection and adaptation as the be-all of evolution, the systematist is indeed left with nothing intrinsic to his or her own material to explain.

### Mayr on Variation and the Polytypic Species Concept

Darwin (1859), Mayr notes (1942, p. 147), never addressed the question of the origin of species in his book bearing that title. To Darwin, the antithesis to the notion of evolution was "fixity of species." And species fixity came to be nearly synonymous with "reality of species." Darwin argued that species were changeable, not immutable. Their seemingly permanent nature is an illusion of the moment. All species looked different in the not-so-distant past and are destined to change further in the not-so-distant future. Species, in this view, are transitory, ephemeral carriers of anatomical characteristics at a particular stage of evolutionary development at any particular moment. To Darwin, species were not concrete, real objects; their "origin," then, hardly required explanation. What did require explanation was the modification of old structures into new, the change of intrinsic features of organisms, which is still a basic definition of the word *evolution* today. The species concept of all strict adherents to the synthesis (see Bock 1979 for a recent and explicitly reductionist view of species) logically requires this viewpoint. The first part of Mayr's 1942 book is a determined attempt to support the notion that species exhibit variation within populations and among populations geographically. The variation is genetically based to a great degree. Moreover, the variation is nearly always adaptive; the sorts of characters used to distinguish species are no different in kind from those seen varying within species. There is, in essence, a continuum from patterns of variation within local populations right up through the kinds of differences among clusters of closely related species. A species, looked upon in this way, is merely a transitory stage of a continuous stream of adaptive differentiation. This view is, of course, perfectly Darwinian and agrees completely with notions of adaptation being developed by both Dobzhansky (1937a, 1941) and Simpson (1944).

To develop his ideas, Mayr first attacked the typology of his fellow systematists—the tendency to treat species as if they do not vary, hence the frequent description of specimens only slightly different from the standard reference (or "type") specimen of a species as a "new" species. Mayr attacks the typological species concept in his first chapter as a forerunner to his argument that species are inherently variable: In the new systematics, "the importance of the species as such is reduced, since most of the actual work is done with subdivisions of the species" (Mayr 1942, p. 7).

Though the deemphasis of the role of species as such pertains to systematics and not to evolutionary theory per se, it sets the tone for much of what follows in the three succeeding chapters: the overriding importance of within- and among-*population* variation, the microanatomy of species. The entire purpose of chapter 1, ostensibly devoted to "methods and principles of systematics," is to show that the change in emphasis within systematics (as perceived by Mayr) is based on a fundamental biological reality; systematics was finally recognizing the nature and significance of geographic variation and coming to see that species are not the monoliths they were once supposed to be. Mayr makes this point explicitly later in the book (1942, p. 108), when he discusses the "new species concept."

In chapters 2 through 4 there is little developed to show how the demonstration of internal variation within species and its gradationally smooth continuation into among-species differences is itself an attack on the internal cohesion of species. But, in the preface to an earlier reprinted edition of his 1942 book, Mayr himself draws attention to it:

In 1942, it was most important to demolish the typologically defined species. The emphasis had to be on the subdivisions of the species, on subspecies and local races. This prepared the way for a treatment of geographic variation and of speciation. The conclusions derived from this approach are now so generally accepted that it has again become important to stress the unity of the genotype in a species, and this is the key concept of the [Mayr] 1963 volume. (Mayr 1964, p. x)

The implicit and only retrospectively recognized dilemma is not resolved until later chapters (5 and 6), where gaps between species are admitted and discussed. Species are variable, indeed, spread out over their geographic ranges, but they are nevertheless reproductively isolated and generally morphologically distinguishable from other species.

Mayr (1942, pp. 75, 86) repeatedly says that he is not claiming that all variable characters are either genetically based or of adaptive significance. But it is clear that he feels most variation is adaptively generated, though adaptive variation per se is not addressed until chapter 4, the second of two chapters on geographic variation. He notes (p. 57, chapter 3) that "all the characters which have been described as good species differences have been found to be subject to geographic variation" and (p. 59) "geographic variation not merely helps in producing differences, but [also] many of these differences, particularly those affecting physiological and ecological characters, are potential isolating mechanisms, which may reenforce an actual discontinuity between two isolated populations [see chapter 9]." In other words (p. 59), "geographic isolation is thus capable of producing the two components of speciation, divergence and discontinuity."

That Mayr concludes that such variation is genetically based and frequently adaptive in nature is made abundantly clear in the penultimate paragraph of chapter 3:

First, there is available in nature an almost unlimited supply of various kinds of mutations. Second, the variability within the smallest taxonomic units has the same genetic basis as the differences between the subspecies, species, and higher categories.[1] And third, selection, random gene loss, and similar factors, together with isolation, make it possible to explain species formation on the basis of mutability, without any recourse to Lamarckian forces. (Mayr 1942, p. 70)

Mayr's final chapter (the fourth) before he turns to a consideration of species definitions per se, expressly considers adaptive geographic variation. Acknowledging both accidental differences and neutral polymorphism (Mayr's occasional reference to selective neutrality has a distinctly modern ring), he states (1942, p. 86) that "it should not be assumed that all the differences between populations and species are purely adaptational and that they owe their existence to their superior selective qualities." In what is really a stylistic technique employed many times in the book, he stresses (p. 86) "the point that not all geographic variation is adaptive"—before assaulting the reader with a mountain of evidence and a wealth of detail, all designed to show, in this case, that geographic variation most certainly is usually adaptive.

The bulk of Mayr's treatment of adaptive geographic variation is devoted to a consideration of ecological rules and the nature of clines. In each discussion there is a rather surprising conclusion from one who claims that a complete intergradation in patterns of adaptive variation within and among species is a general rule of nature. For, on page 94, after a still useful review of ecological rules, Mayr says that "these rules apply only to *intraspecific* variation"—that is, they do not describe patterns of interspecific variation.

Similarly, after discussing clines (i.e., character gradients within a species over some or all of its geographic range, assumed to represent local adaptations to a linearly varying environment), Mayr concludes (1942, p. 97, and he italicizes the sentence for emphasis): "*The more clines found in a region, the less active is species formation.*" The reason: speciation requires discontinuities, but clines indicate continuities. There is a dilemma here, one that Mayr resolves in the second half of his book by invoking geographic isolation, not selection, as the major mechanism for the generation of discontinuities. But the effect of his section on adaptive geographic variation is a bit surprising in its deeper implication: in the only two sorts of phenomena explicitly discussed (ecological rules and clines), with examples, to document the adaptive nature of geographic variation, the relevance to among-species differences (in the case of the ecological rules) and to the origin of species (discontinuities, the case of clines) is expressly denied. Far from showing that among-species differences arise as a smooth continuum of adaptive change from within-species differences, Mayr's actual examples show just the opposite and, perhaps even more interestingly, Mayr is aware of it, as his use of italics for emphasis in each of these quoted passages would seem to imply.

So there is some equivocation. Mayr is saying in his first hundred pages that variation exists, that it is adaptive, and that within-species variation is the same in character as among-species variation; hence the latter springs from the former. He comes close in places to denying that species are anything more than ephemeral collections of populations, intermediate in some loose sense between populations on the one hand and genera on the other. But at the same time he hints at an alternative theme that species are separated from each other by the "bridgeless gaps" that figure so prominently in the remainder of the book. Selection can exert pressure on but cannot pull apart these things called species—a decisive position on an issue that troubled Dobzhansky so much (see chapter 2 herein) and a conclusion diametrically opposed in essence to Dobzhansky's solution to the problem. I will discuss further below Mayr's view that selection itself plays no role in the establishment of reproductive isolation. This attitude explains, among other things, Mayr's almost legendary abhorrence of sympatric speciation.

### More on Species

To Mayr in 1942, it seemed a paradox for the evolutionist to attempt to define species, to try

to establish a fixed stage in the evolutionary stream. If there is evolution in the true sense of the word, as against catastrophism or creation, we should find all kinds of species—incipient species, mature species, and incipient genera, as well as all intermediate conditions. To define the middle stage of this series perfectly, so that every taxonomic unit can be certified with confidence as to whether or not it is a species, is just as impossible as to define the middle stage in the life of man, mature man, so well that every single human male can be identified as boy, mature man, or old man. It is therefore obvious that every species definition can only be an approach and should be considered with some tolerance. On the other hand, the question: "How do species originate?" cannot be discussed until we have formed some idea as to what a species is. (Mayr 1942, p. 114)

Thus does Mayr express the dilemma, this contrast between seeing morphological divergence as a continuum versus the discrete nature, the very existence, of species, the origins of which are to be construed as fair game for scientific inquiry. The problem, essentially the same one already apparent in his first hundred pages, is stated in a paragraph immediately following Mayr's criticism of Kleinschmidt and Goldschmidt for themselves failing to define species—in Mayr's view a fatal flaw in their argument that a species is "separated from other species by bridgeless gaps as if it had come into being by a separate act of creation" (Mayr 1942, p. 114). For, as Mayr claims, this was the paradoxical conclusion of these men who had been among the key figures, with Rensch and a few others repeatedly cited by Mayr, in the development of contemporary concepts of geographic variation. Mayr writes (p. 114): "As the new polytypic species concept began to assert itself a certain pessimism seemed to be associated with it." All those separate species that seemed to intergrade now were viewed as widespread polytypic species, and the evidence of morphological intergradation among species was diminished, not enhanced. Mayr chides his colleagues for this defeatest attitude but nonetheless does not explicitly attempt to refute it. And, as we have seen, while his first hundred pages purport to refute this notion, it was only with a mixed success of which the author seems aware.

Moreover, though he does criticize Kleinschmidt and Goldschmidt, Mayr himself adopts the phrase *bridgeless gaps* throughout the rest of his book. His view of the polytypic species seems exactly the same as Goldschmidt's, as is evident from his famous definition, the short version of which is still memorized by students: "Species are groups of actually or potentially interbreeding natural populations, which are reproductively isolated from other such groups" (Mayr, 1942, p. 120). The bridgeless gap, in this case, is reproductive isolation; coherency within species is maintained by a plexus of parental ancestry and descent, and the very process of reproductive interaction that binds a species together keeps it apart from other species. This is the prevailing view of species in evolutionary biology today. It has been criticized (e.g., Simpson 1961; Wiley 1978) for its alleged two-dimensional, ahistorical nature, and on epistemological grounds by numerical taxonomists (e.g., Sokal and Crovello 1970) and some cladists concerned with the practical nature of species recognition.

With his definition of species, Mayr is, of course, confronting the ontological status of these kinds of taxa. He remarks that Dobzhansky's definition is "an excellent description of the process of speciation, but not a species definition. A species is not a stage of a process, but the result of a process" (Mayr 1942, p. 119). And Mayr's own analogy between stages in the life of a man and stages of development of the products of the evolutionary process suggests that he does see species partly as a mere transient stage in the evolutionary stream. Yet when he comes to define species, he does so in such a way as to suggest that they are in some important sense real entities after all, so that he can treat them as actual objects and proceed to discuss their origins.

The ontological status of species—specifically the option of considering species as classes or as individuals (following the analyses of Ghiselin 1974a and Hull 1976)—is perhaps the most crucial issue so far broached that bears on the success of the synthesis as an all-embracing theory of evolution. Viewing species as individuals opens the door to a more fully explicit and completely hierarchical view of nature's organization—and to a concomitantly hierarchically based theory of life's evolution. Part of Mayr's discussion sees species as individuals, which is not the predominant notion of species within the synthesis—nor the only view Mayr espouses within the pages of *Systematics and the Origin of Species*. If it is true that Mayr did not invent the notion that species are real entities, it is nonetheless indubitable that the underlying basis for all the various recent approaches to looking at species as entities, or individuals, rather than as classes of individuals stems from Mayr's species concept, first published in 1940 but established and recognized primarily upon publication of his 1942 book.

Mayr directly addresses the ontological question of the reality of species in his chapter 7, "The Species in Evolution." Acknowledging (p. 147) that many biologists regard species as mere abstractions, considering "the individual [i.e. organisms] the only unit in nature which possesses any reality," Mayr asserts (p. 148) that "we come to the conclusion that such a unit is objective, or real, if it is delimited against other units by fixed borders, by definite gaps."

A few pages later, Mayr reveals his ambivalence on the ontological status of species on facing pages. His discussion is actually couched epistemologically: How do we classify allopatric taxa? In particular, how can we tell, in any given case, whether we have allopatric subspecies within a single polytypic species, or, instead, a series of allopatric species? Similar problems arise with allochronic (i.e., fossil) taxa. But in raising these methodological problems, Mayr clearly reveals his views on the ontological issue: To what extent are species real entities in nature? In his discussion of these problems, we find one of many examples where an epistemologically murky area, so troublesome to the museum taxonomist, becomes (according to Mayr) an evolutionist's delight: the decision about whether to call such forms species or subspecies is often "arbitrary and subjective" (p. 152). But, he continues,

this is only natural, since we cannot accurately measure to what extent reproductive isolation has already evolved. In fact, such cases are logical postulates, if the divergence of isolated populations is one of the important means of species formation. A species evolves if an interbreeding array of forms breaks up into two or more reproductively isolated arrays, to use Dobzhansky's terminology. If we look at a large number of such arrays (that is species), it is only natural that we should find a few that are just going through this process of breaking up. This does not invalidate the reality of these arrays; just as the *Paramecium* "individual" is a perfectly real and objective concept, we find in most cultures some individuals that are either conjugating or dividing. (Mayr 1942, p. 152)

Thus, Mayr's position on the nature of species finally becomes quite clear in his chapter 7, after the two chapters explicitly devoted to species definitions and characterizations. When they are sympatric, species are individuals separated from others by unbridgeable gaps. There is no question of their ontological status, and in practice there is usually no trouble in recognizing them. In the allopatric case, the situation becomes epistemologically messy, but the ontology remains clear: if a *Paramecium*, an individual before fission, becomes two individuals after fission, the blurry situation during the brief interval of actual fission is of no material importance.

Allochronically, however, Mayr believes that species give up the ghost ontologically. In a passage on the very next page after his *Paramecium* analogy with species, Mayr makes the usual claim that, were the fossil record perfect, we would see no discrete species. He professes thanks that the fossil record is sufficiently full of gaps to afford us the arbitrary criteria to recognize separate species within fossil lineages:

Hitherto we have spoken only of the delimitation of contemporary (synchronic) species. The delimitation of species which do not belong to the same time level (allochronic species) is difficult. In fact, it would be completely impossible if the fossil record were complete. The species of each period are the descendants of the species of the previous period and the ancestors of those of the next period. The change is slight and gradual and should, at least theoretically, not permit the delimitation of definite species. In practice, the fossil record is fragmentary, and the gaps in our knowledge make convenient gaps between the "species." (Mayr 1942, p. 153)

Mayr goes on to say that the paleontologist treating one of the few cases of uninterrupted fossil series follows the reasoning of the taxonomist confronting continuously intergrading sequences: "He breaks them up for convenience" (p. 154). But, though such continuity in space on a single time plane poses no threat to the reality of species, Mayr sees no comparable status for such evolving lineages through time. To Mayr in 1942, species are real, discrete individuals in space at any one point in time. Through time, species lose their discreteness, their identity, their individuality. They become stages in a stream, arbitrarily delineated segments of a continuum. They become classes.

## The Bridgeless Gaps

It is in the early pages of his chapter 7, the first of three chapters devoted to speciation per se, that Mayr turns to the subject of discontinuities between species and the nature of their origin. This is where he points out that Darwin's title was a misnomer. And it is here (p. 148) that Mayr tells us that gaps give species reality. Mayr then discusses these gaps, asking (p. 148): "Do such gaps exist and how complete are they?" Mayr's gaps are morphological as well as reproductive; sometimes they are primarily reproductive, and at other times they are clearly morphological in nature.

To Mayr, such gaps are clearcut among sympatric species and gradational among allopatric species—leading to the discussion already cited
on the ontological and epistemological status of species. Incidentally, it is
here (pp. 148–49) that the terms *sympatric* and *allopatric* are first defined
(though Mayr cites Poulton as the coiner of *sympatric*). The gaps are
bridgeless only in the sympatric case—not in the allopatric case, where
they are gradational. Mayr (1942, p. 149) chastises Goldschmidt, who,
Mayr feels, extrapolates too far by extending his evidence of the bridgeless gaps among sympatric species to embrace the allopatric situation as
well.

Thus, the situation is again equivocal. At one time the *Paramecium* metaphor of individuality, despite allopatric cases where budding is occurring, upholds the notion of species as discrete, real individuals. But
throughout the remainder of chapter 7, Mayr speaks of stages of
speciation:

That speciation is not an abrupt, but a gradual and continuous process is proven
by the fact that we find in nature every imaginable level of speciation, ranging
from an almost uniform species at one extreme to one in which isolated populations have diverged to such a degree that they can be considered equally well as
separate, good species at the other extreme. (Mayr 1942, p. 159)

Mayr's resolution of the problem of the origin of gaps—be they reproductive or morphological—is well known: geographic isolation leads to
reproductive isolation, which in turn allows gaps to appear among morphologically differentiated populations of the ancestral species. Mayr
writes:

*A new species develops if a population which has become geographically isolated
from its parental species acquires during this period of isolation characters which
promote or guarantee reproductive isolation when the external barriers break down.*
(Mayr 1942, p. 155, italics in original)

Mayr argues that the establishment of discontinuities is a continuous
process, with large gaps occurring between species, smaller gaps between
subspecies, and still lesser gaps between populations. "Of course, if the
populations are distributed as a complete *continuum*, there are no gaps.
But with the least isolation, the first minor gaps will appear" (Mayr 1942,
p. 159). In other words, morphological diversity (established through
adaptive geographic variation) plus isolation yields discontinuity. Mayr's
fig. 16, depicting in graphic form his notion of the stages of speciation, is
reproduced here as fig. 3.1.

Mayr then proceeds to document what he claims is "conclusive proof
for the existence of geographic speciation: If an isolated population of a
species remains long enough in this isolation, it may acquire biological
isolating mechanisms which permit it, after the breakdown of the isolating
barrier, to exist as a separate species within the range of the parental species. The reproductive isolation, which originally was maintained by the
extrinsic means of a geographic barrier, is being replaced during this iso-

*Stage 1.* A uniform species with a large range.

Followed by:
*Process 1.* Differentiation into subspecies

Resulting in:
*Stage 2.* A geographically variable species with a more or less continuous array of similar subspecies (2a all subspecies are slight, 2b some are pronounced.

Followed by:
*Process 2.* Isolating action of geographic barriers between some of the populations; also development of isolating mechanisms in the isolated and differentiating subspecies.

Resulting in:
*Stage 3.* A geographically variable species with many subspecies completely isolated, particularly near the borders of the range, and some of them morphologically as different as good species.

Followed by:
*Process 3.* Expansion of range of such isolated populations into the territory of the representative forms.

Resulting in either:
*Stage 4.* Noncrossing, that is, new species with restricted range.

Or:
*Stage 5.* Interbreeding, that is, the establishment of a hybrid zone (zone of secondary intergradation).

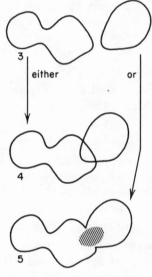

Figure 3.1.   Mayr's fig. 16 (1942, p. 160): Stages of speciation.

lation by intrinsic isolating barriers. One species has developed into two" (Mayr 1942, p. 185).

Mayr's basic conception of allopatric speciation—stemming from the earlier work of von Buch, Wagner, Romanes, and others—has stood the test of time well. A review of speciation theory by Bush (1975) and a book by White (1978) reveal that, in addition to the allopatric scheme Mayr espoused, various other notions of parapatric and sympatric speciation have gained a measure of acceptance in recent years. Mayr is famous as much for his intrepid opposition to notions of what he calls (1942) "non-geographic speciation" as for his promulgation of the theory of allopatric speciation. Admitting that parasitic organisms might qualify as candidates for sympatric speciation, Mayr (1942, chapter 8) does his best to stamp out the notion of anything but pure geographic speciation. The reason for this position is obvious: neither selection nor any other genetic mechanism, in his view, can create reproductive discontinuity, no matter how much morphological divergence can occur within a local population. And species are reproductively isolated from one another. The one solution Mayr could accept was geographic isolation.

The point is important because it shows that, despite the fact that Mayr has morphology in mind most often when speaking of interspecific gaps and despite the fact that most of the book is devoted to morphological differences within and between species, Mayr is also deeply concerned with the origin of reproductive discontinuity—the sine qua non of all gaps. Reproductive gaps may exist with virtually no correlative morphological gaps, but in no case are there morphological gaps without reproductive gaps.

A closely related point deserves attention: Mayr treats sibling species (two closely similar sympatric species usually, though not necessarily, quite closely related) in two separate places in the book. They are discussed in chapter 7 ("The Species in Evolution"), where they "demonstrate clearly that the reality of a species has nothing to do with the degree of its distinctness" (p. 151). Sibling species reappear in chapter 8, where Mayr claims (p. 208) that he is not surprised that they exist; in fact, he says, it is surprising that "the acquisition of reproductive isolation is so often combined with the acquisition of distinct morphological novelties." This is the converse of his observation on page 288, where he says: "We can have much divergent evolution without the origin of new species and considerable speciation without much evolutionary divergence." This is the theme, quite popular in present-day evolutionary biology, of the quasi-independence of morphological evolution and the origin of new taxa (species). In the particular case of sibling species, Mayr is saying that they ought to be more common than they apparently are because of this "decoupling," and he makes this point even though the sympatric coexistence of nearly identical and generally closely related species would seem at first view to pose an embarrassing conundrum to such an ardent exponent of allopatric speciation (a difficulty not acknowledged in this book, incidentally). But Mayr's treatment of sibling species belies a deeper interest: reproductive gaps are more fundamental than morphological gaps in the evolutionary process.

It is in his chapter 9 that Mayr comes to grips with the basics of the speciation process; the factors that govern the establishment of both divergence and discontinuity. He starts off as follows:

We may classify these factors as (1) those that either produce or eliminate discontinuities and (2) those that promote or impede divergence. The latter may be subdivided further into adaptive (selection) and nonadaptive factors (see Huxley 1941).(Mayr 1942, p. 216)

This is a logical scheme indeed, following Mayr's explicit conception (pp. 23, 158, and many other places throughout the text) of the two basic ingredients of speciation. But Mayr does not follow this scheme, preferring instead to "treat divergence and discontinuity together and to select instead a different scheme of classification" (p. 216).

The remainder of the chapter follows his classification of the factors governing speciation; the two main subdivisions concern "internal" and "external" factors. The classification is by no means clear; extinction, for example, is treated under internal factors. Mayr—with interestingly little explicit citation of Dobzhansky's (1937a, 1941) pioneering discussion of isolating mechanisms—sees two kinds of "isolating factors": "geographic and reproductive barriers." On the one hand, Mayr says (p. 226), speciation cannot occur if "biological" isolating mechanisms do not develop during geographic isolation. Yet such biological isolating mechanisms cannot develop without geographic isolation.

Mayr rearranges the classification of isolating mechanisms (presumably Dobzhanky's) by adding:

Isolation, according to this classification, may be accomplished because the potential mates do not meet (*restriction of random dispersal*), or because they do not mate even if they meet (*restriction of random mating*), or because they do not produce their normal quota of offspring even though they mate (*reduction of fertility and related phenomena*). (Mayr 1942, p. 226)

The first case is geographic isolation *per se*; the other two are Mayr's biological isolating mechanisms, including ecological, ethological, mechanical, and genetic and physiological factors (listed in Mayr 1942, pp. 247–8).

Mayr directly confronts the role of selection as a factor in speciation in a brief discussion beginning on his page 270. Alluding to his earlier discussion on the adaptive origin of morphological diversity, Mayr focuses here only on predation and competition, concluding (as he has done all along) that selection is an important factor (and the only important deterministic one) in the development of morphological diversity, but it plays a negligible role in the establishment of discontinuities. His discussion of the problem is a mere two and a half pages long and, again, makes no reference to Dobzhansky's agonizing over the role selection might play in the development of reproductive isolation. Mayr, in particular, pays no heed to Dobzhansky's reinforcement model in this section, though earlier in his book the notion is discussed:

The speciation process does not need to be completed during this isolation. Dobzhansky . . . has pointed out that selective mating in a zone of contact of two formerly separated incipient species (zone of secondary intergradation . . .) may play an important role. The two incipient species must be sufficiently distinct, so that the hybrid offspring of mixed matings has discordant (unbalanced) gene patterns; in other words, the individuals produced in such matings must have a reduced viability and survival value. (Mayr 1942, p. 157)

Mayr then quotes Dobzhansky's model. The quotation of Dobzhansky ends: "Consequently, natural selection will favor the spread and establishment of the mutant condition . . ." and then continues in Mayr's own words:

until only conspecific pairs are formed or, in other words, until complete discontinuity (a bridgeless gap) has developed between the two species. Dobzhansky presents a plausible case, and we agree that such a selective process may help to complete the establishment of discontinuity, in those cases in which some interbreeding has taken place between incipient species.

The question is, however, whether or not this is the only way by which reproductive isolation can be established. (Mayr 1942, p. 157)

And Mayr makes it abundantly clear that the answer is no.

As we have seen, Dobzhansky (1937a, 1941) argued in favor of the adaptive advantage of "reproductive isolation." Selection, of course, is implicated in the origin of reproductive isolation because that isolation is adaptively advantageous, keeping a species firmly established on an adaptive peak and preventing the wholesale development of adaptively disadvantageous gene combinations. Isolation prevents miscegenation. Paterson (1978, 1980, 1981, 1982) has argued against this entire viewpoint, claiming in effect that isolating mechanisms between any two species are nothing but fortuitous divergences in what he calls "specific mate recognition systems." Selection acts solely to enable mates within disjunct populations to recognize one another; it does not act to keep two populations from interbreeding. But it is also clear that Mayr, like Dobzhansky in some contexts, saw isolating mechanisms as "specific recognition marks. Their primary function seems to facilitate the meeting and recognition of conspecific individuals and to prevent hybridization between different species" (Mayr 1942, p. 254).

## Mayr on Macroevolution

Mayr concludes his book (1942, chapter 10) with a discussion of "the higher categories and evolution." The bulk of his discussion dwells on "macrotaxonomy" (see Eldredge 1982a for a discussion). But he devotes his final eight pages to "macroevolution." Mayr (p. 291) construes macroevolution as "the development of major evolutionary trends, the origin of higher categories, the development of new organic systems—in short, evolutionary processes that require long periods of time and concern the higher systematic categories (supraspecific evolution). There is only a difference of degree, not one of kind, between the two classes of phenomena. They gradually merge into each other and it is only for practical reasons that they are kept separate."

Alluding to epistemological differences among the fields of paleontology, taxonomy, and genetics, Mayr emphasizes his conviction (p. 292) that just because paleontologists have erred in some of the explanations of their data is no reason to ignore the data themselves.

Mayr then cites Dobzhansky and Rensch to the effect that "there are already many genetic phenomena known which deprive the macroevolutionary processes of much of their former mysteriousness." Among these he lists (pp. 292–93): "1. the smallness and frequency of mutations . . . 2. the multiple (pleiotropic) effect of genes . . . 3. the multiple genic basis of a single trait . . . 4. many mutations reversible . . . 5. the growth of individual organs and structures often a function of the growth of the whole organism . . . 6. the speed of evolution a matter as much or more of extrinsic as of intrinsic factors; the influence of isolation, swamping, population size, accidents of sampling . . . 7. selection a powerful evolutionary factor . . . , and 8. the evolutionary specialization of one organ system compensated for by the reduction of others." All these topics and assertions are already familar from Dobzhansky (1937a, 1941; see chapter 2 herein), with the exception of items 5 and 8.

Mayr claims (p. 293) his list supplies "only some of the known genetic and evolutionary factors which help to explain macroevolutionary 'laws' and phenomena," which include orthogenesis, size increase in phylogenetic lines, convergence, specialization, irreversibility, "the principle of harmonious organic reconstruction" (p. 295), and "the principle of conservative characters" (which Mayr admits on his p. 296 is something of a puzzle).

Mayr closes his book with the "principle of the instantaneous origin of new types"—the claim (Mayr cites Schindewolf 1936) "that all or nearly all the major types of animals have appeared on the earth in a more or less finished form, with the links to the presumable ancestors missing" (p. 296). Mayr (p. 297) is in close agreement with Dobzhansky's position on the nature and cause of systemic gaps in the fossil record—a theme that provides the focal point to Simpson's (1944) entire book. Mayr agrees that saltation[2] is not necessary to explain the origin of taxa of higher categorical rank.

Mayr raises yet another issue from paleontology (p. 297): "the different rates of evolution in different groups or in different periods of the same group." Here, in one of his few allusions to adaptive peaks, (and the only one to use the metaphor in a truly extended sense), Mayr speculates (p. 297) that groups "searching for a new 'adaptive peak' may undergo rapid evolution, but as soon as this peak has been reached evolution may begin to stagnate." This is the theme of rapid exploitation of new possibilities, a theme emphasized, for instance, particularly by Simpson (1944, but especially 1953 and 1959a) as well as others (e.g., Eldredge and Tattersall 1982).

Thus, it comes as no surprise whatsoever to read Mayr's concluding paragraph:

In conclusion we may say that all the available evidence indicates that the origin of the higher categories is a process which is nothing but an extrapolation of speciation. All the processes and phenomena of macroevolution and the origin of the higher categories can be traced back to intraspecific variation, even though the first steps of such processes are usually very minute. (Mayr 1942, p. 298)

Again we see that the bridgeless gaps between species, the very existence of species themselves, had no special significance for Mayr in 1942—at least when it came time to assess the role of species in evolution. When thinking of macroevolution, Mayr saw the problem simply as transformational, the modification of genotype-phenotype according to the principles of genetics to a sufficiently great degree to produce the various macroevolutionary phenomena he cites. As Mayr himself has pointed out, and as Eldredge and Cracraft (1980, p. 275) have also seen, later in his career Mayr came to see species as such as playing a far more active role in the production of large-scale evolutionary patterns. Thus, the duality of his view on the ontological status of species—as portrayed so vividly on the facing pages 151 and 152 of *Systematics and the Origin of Species*—is critical to understanding both Mayr's basic stance in this book and the later development of his thought. In 1942 he scoffed at people who claimed that species were not real, yet he managed to see species through time as transient stages in the stream of evolution within the very same book. In discussing species' origins, the former view prevailed: Species must exist, else why do we need a theory to explain their origins? Yet, through time, species are classes, as Mayr's discussion of macroevolution so abundantly confirms.

One final comment before passing on to George Gaylord Simpson's *Tempo and Mode in Evolution*: I began this discussion with Mayr's recently published list of eight items he claims systematics contributed directly to the newly formed synthesis. All these items are prominently displayed in his 1942 book, so much so that they read like a synopsis of his argument: population thinking, the immense variability of populations, the gradualness of evolution, the genetic nature of gradual evolution, geographic speciation, the adaptive nature of observed variation, belief in the importance of natural selection, and the gradual nature of macroevolution.

## TEMPO AND MODE IN EVOLUTION

From Mayr's 1942 treatment of the evolutionary process from the standpoint of systematics, we turn now to the work of paleontologist George Gaylord Simpson, Mayr's colleague at the American Museum of Natural History. I have cited Simpson's opening words at the head of my own first chapter simply because it is as true today as it was in 1944 that the data of a broad spectrum of biologial disciplines are all directly relevant to a thorough undertanding of the evolutionary process. Yet Simpson sought to integrate the data and especially the ideas of his own discipline—paleontology—with genetics in particular: "The attempted synthesis of paleontology and genetics, an essential part of the present study, may be particularly surprising and possibly hazardous" (Simpson 1944, p. xv). While his venture was presumably neither particularly surprising nor hazardous, it is worth pointing out at the outset of this analysis of Simpson's work that his book is in many ways more comparable with Dobzhansky's (1937a, 1941) than it is with Mayr's (1942)—this depite the fact that paleontologists, too, are systematists in every bit as full a sense as are students of living organisms. Where Mayr eschews genetics in favor of exploration of the adaptive nature of phenotypic variation within species, Simpson looks for particular combinations of genetic determinants—the variables so much discussed by Fisher, Wright, Haldane, and, of course, Dobzhansky—that might account directly for the patterns he sees in the fossil record. Thus, it is in a sense surprising to find that both Mayr's and Simpson's books can more appropriately be compared with Dobzhansky's work than with each other.

### Simpson's Evolutionary Theory

*Tempo and Mode in Evolution* is far more than a mere consistency argument that paleontological patterns can be explained by neo-Darwinian principles—the sort of consistency argument (as Gould 1980a points out) that is so typical of the writings of the synthesis. It is, in addition, a dense and minutely reasoned essay in which hardly a word is wasted throughout. It is a very detailed particular theory of evolution, in some respects highly idiosyncratic. Some of the idiosyncrasies are uniquely Simpson's, but some are those of the discipline of paleontology itself. For this book, far more than its successor (*The Major Features of Evolution*, Simpson 1953), simultaneously seeks to integrate paleontology with the principles of genetics and to reserve for paleontology a unique place in evolutionary biology. In the early 1940s, Simpson was eager to keep more of the job of evolutionary theorizing away from the geneticists than his subsequent writings seem to allow. In a profession where the intellectual tradition of the 1930s held the work of geneticists to be irrelevant to understanding their data (see Gould 1980a; Wright 1945), Simpson tried to demonstrate such relevancy, but he also tried to maintain that at least some paleontological phenomena were different in kind—and not merely in degree— from the ordinary sorts of phenomena studied by geneticists and systematists working on populations, subspecies, and species in the laboratory and in the wild. The basic ingredients ("determinants," as Simpson called them) were the same, but they were combined in different forms to explain qualitatively different patterns.

The current tendency to see the synthesis as a thoroughgoing reductionist statement (e.g., Eldredge and Cracraft 1980, chapter 6; Gould 1982b) in part overlooks the more complicated—and to some extent explicitly hierarchical—formulations of the early works. A case in point is Simpson's *Tempo and Mode*. I will argue here that Simpson's reductionism in 1944 is a far more complex affair than its subsequent, more thoroughgoing manifestations. In most respects, *Tempo and Mode* fits the reductionist mode to a tee. In chapter 2, the book's longest, entitled simply "Determinants of Evolution," Simpson lists variability, rate and character of mutations, length of generations, size of populations, and natural selection—all items from the repertoire of genetics. That Simpson saw no problems or special difficulties in extrapolating such phenomena and considerations to the paleontological scale of millions of years is abundantly evident throughout the book. For example, when discussing the common charge that to qualify as a valid principle natural selection should faithfully track every environmental change automatically, he speaks of (p. 163) "opponents of the theory of orthoselection (or of selection in general)"—orthoselection to Simpson being natural selection maintained sufficiently long (and not necessarily rigidly unidirectional or constant in intensity) to bring it into the realm of paleontological phenomena. The phrase quoted here suggests the flavor of process continuity that Simpson sees (indeed assumes) from the generation-by-generation levels of the geneticists all the way up to the large-scale patterns of the fossil record. This almost casual style of extrapolation, of course, is required by a reductionist approach to evolutionary theory.

The ontological status of species and taxa of higher rank—specifically the notion that taxa are individuals, not classes—is the main objection usually cited to the reduction of macroevolution to the principles of genetics (e.g., see Eldredge 1982b). The notion of species as spatiotemporally discrete individuals, and of taxa of higher categorical rank as merely monophyletic groups of such entities (species), is seen as a challenge to the notion that within-species generation-by-generation change is necessary and sufficient to explain among-species differences—another way of stating the reductionist thesis of the synthesis. From this current perspective, it is important to examine Simpson's view of the nature of taxa in 1944.[3]

Simpson's view of taxa in *Tempo and Mode* is inextricably bound up with his view of their origins. Taxa are alike in kind throughout the taxonomic hierarchy from phyla on down through subspecies. That all taxa are merely arbitrarily delineated segments of progressive change within phyletic lines is revealed in many passages thoughout his text.

Taxa, to Simpson, are stages of the progressive divergence and development of anatomical change in evolution. Even quantum evolution—which, I shall argue, is the central theme of this book—provides no exception. Simpson says that quantum evolution is mostly involved with the origins of taxa of higher categorical rank—the "mega-evolution" of his chapter 3. He writes:

From another point of view mega-evolution is, according to this theory, only the sum of a long, continuous series of changes that can be divided taxonomically into horizontal phyletic subdivisions of any size, including subspecies. When a new class, for instance, arises, its transitional stages can be divided into subspecies, species, and so forth, on morphological or genetic criteria. It differs from its ancestry first to a degree that would be ranked as subspecific among contemporaneous groups, then as specific, then generic, and so forth, but the nominal subspecies in this process of successive, cumulative change arise by one particular sort of sub-speciation, which is not the most common process of, for instance, geographic sub-speciation. (Simpson 1944, p. 124)

The notion that species are spatiotemporally discrete entities stems, to a great degree and as we have seen, from the arguments of spatial (i.e., not temporal) discreteness developed by Mayr (1942). Simpson expresses his view on the origins of such discontinuities: "The crucial point in pop-ulation differentiation is that at which discontinuity appears between parts of what was previously a continuous population" (Simpson 1944, p. 97).

But one page later, in a discussion of the "development of the discon-tinuity between *Merychippus* and *Hypohippus*," Simpson insists

that the horizontal discontinuity between species, genera, and at least the next higher categories can arise by a process that is continuous vertically and that new types on these taxonomic levels often arise gradually at rates and in ways that are comparable to some sorts of subspecific differentiation and have greater results only because they have had longer duration. (Simpson 1944, pp. 98–99).

And, in his discussion of speciation (p. 199), Simpson continues the theme of gradual divergence underlying the differentiation of taxa. Split-ting (the origin of reproductive isolation), the formation of two or more species from a single ancestor, is merely a special case of the pattern of phyletic change that leads to taxa of higher rank. Simpson here acknowl-edges the "great evolutionary significance" of genetic isolation yet claims that speciation leads nowhere, generally remaining within one adaptive zone. Mayr and Simpson agree that species are discrete entities in space but not in time; however, to Mayr the nature and mode of origin of this discreteness is critical, while to Simpson, despite his disclaimer, it is of no general significance. Simpson's view of the nature of taxa purely as arbi-trary segments of evolving continua—phyletic lines—is abundantly clear and consistent with his basically reductionist outlook.

What, then, is there about this book that deviates from the reductionist theme? The answer is pattern. Simpson claims that paleontologists see patterns unique in style, qualitatively different from the patterns seen in the data of geneticists and systematists. These patterns require unique and qualitatively different explanations—involving the same ingredients as geneticists use in explaining their own patterns, to be sure. But new the-ory involving these genetic ingredients must be invented to explain the paleontological patterns.

In spite of the rich assortment of paleontological phenomena to be found in Simpson's book, he is mostly concerned with one big issue: the problem of gaps between taxa. Paleontologists, he observes, have long known that new taxa at all levels (species through phyla) usually appear suddenly in the fossil record, without smoothly gradational annectant forms between ancestor and descendant. As he remarks in his chapter 3, both geneticists such as Richard Goldschmidt and paleontologists such as Otto Schindewolf claim that the record is to be taken literally; hence their (various) notions of direct, sudden, single-step development of one taxon from another. Others claim an imperfection of the record—the standard cry of all good Darwinians from the master himself through such mentors of Simpson as W. D. Matthew. Simpson says the truth lies somewhere in the middle. Specifically, he argues that gaps between taxa of relatively low rank are almost always the product of a missing record. But gaps between taxa of higher rank are not artifacts of the gappy fossil record; they are real phenomena requiring special explanation. These gaps require a thesis involving high rates of evolution. And these rates are not merely the upper end of a normal distribution of rates. They are instead a special class of rates, set apart from the class of normal (intermediate) rates and the third, equally discrete distribution of abnormally low rates. These three classes of evolutionary rates, according to Simpson, have their own typical combinations of "determinants"—size of population, role of mutation and selection, and so forth. They are also involved in peculiar ways with notions of adaptation, and when they are combined with the latter a set of three quasi-independent modes can be inferred: speciation (changed to "splitting" in Simpson 1953), phyletic evolution, and quantum evolution. Simpson himself claims (1944, p. 197) that mode is actually inferred from tempo, referring to "the three modes that are to be distinguished, largely on the basis of all the discussion of tempo and of other evolutionary factors of previous pages."

But it is more accurate perhaps to view *Tempo and Mode* as one complex and tightly knit argument, essentially ad hoc in nature, designed to explain the origin of paleontological gaps between taxa. The reader is led from page 1 of this book to the inevitable conclusion that quantum evolution is the mode of origin of most higher "categories," and by its very nature quantum evolution would produce a pattern seldom if ever directly recorded in the fossil record. It would, instead, produce the gaps we do see. The following analysis of the contents of *Tempo and Mode in Evolution* is geared to demonstrate the structure—and pervasiveness—of this argument.

## Simpson's Argument: The Building of the Theory of Quantum Evolution

Chapter 1 starts out with a bang; the reader is immediately immersed in some arresting details of horse evolution. Saying (Simpson 1944, p. 3) that the question of how fast animals evolve in nature is the first thing geneticists ask of paleontologists, Simpson plunges right into a detailed and sophisticated discussion of evolutionary rates.[4] The entire first chapter is devoted to a demonstration of the fact that paleontologists can study evolutionary rates. First focusing on morphologic rates in equids, Simpson (p. 12) adduces four generalities: (1) rates of change may be causally dependent among characters or wholly independent; (2) more importantly, the rate of change in a character is itself subject to change through time: (3) characters may change independently: and (4) upon divergence of a lineage into two separate phyla, the rate of change of homologous characters is apt to differ. Unlike all subsequent chapters, this one has no succinct summary, and the only general rules are those seen early on (p. 12) for horses: rates of evolution within and among homologous traits may vary enormously within and among lineages. Evolutionary rates do not by any means proceed in lockstep fashion. The data and attendant conclusions are as valid now as they were in 1944, and if they seem less vividly important now, it is in no small measure thanks to this original exposition.

It is here, in his chapter 1, that Simpson first uses taxonomic survivorship curves (so tantalizingly similar to survivorship curves for *Drosophila* individuals, which Simpson acknowledges on p. 26 but steers clear of) as an indicator of gross rates of evolution within taxa of higher categorical rank. Survivorship curves play a fundamental role in his chapter 4, in Simpson's argument that there exist no fewer than three discrete classes of evolutionary tempos. In chapter 1, he merely makes them familiar and shows that among major groups of Metazoa there are great differences in evolutionary rates as seen in the relative survivorship of their constituent genera. His contrasted example—clams versus mammals—is familiar to all subsequent investigators of gross evolutionary rates; the chapter is so successful that, not only are paleontologists still compiling rates forty years after Simpson's example, but they are still actively arguing whether the distinction Simpson claimed—boiling down to a faster rate of appearance of genera in land Carnivora versus marine Bivalvia—is valid (e.g., see Stanley 1973; Van Valen 1976).

Thus, Simpson's chapter 1 tells us that rates of evolution can be measured in various ways and that they can vary within and among different lineages in a complex fashion. No other generalizations are adduced. It is interesting that a whole style of doing theoretical paleontology comes from this essay and its revisions (e.g., Simpson 1953)—including the use of data (e.g., for clams) taken from the literature. But, for the argument of the book—that certain aspects of evolution known only to paleontologists necessitate a particular conceptualization about evolutionary rates—the two points Simpson wanted to convey in this chapter are that (1) evolutionary rates can be measured in an acceptably scientific fashion, and (2) rates vary enormously. These are modest claims when compared with later conclusions as the argument heats up.

In his chapter 2, Simpson turns to substantive issues that will bear directly on his thesis of the nature of the evolutionary process. It is here that he launches into a deliberate review of the "determinants" of evolution which are the special property of genetics: "variability, rate of mutation, character of mutations, length of generations, size of populations, and natural selection" (1944, p. 30). For each topic, Simpson summarizes contemporary thought and, through example, discusses the connection between each topic and paleontological data.

Tackling variability first, Simpson does not question its importance as the basis for the operation of natural selection. Indeed, his first major point in this section is more of a statement on the nature of selection than on variability per se:

From every point of view there is an essential difference between variation of individuals within a group and variation between groups, but the two are often inadequately distinguished. Natural selection, for instance, acts on both, but its action on intergroup variation can produce nothing new; it is purely an eliminating, not an originating, force. Despite its critics, the actions of natural selection on intragroup (or interindividual) variation is essentially an originating force: it produces definitely new sorts of groups (populations), and the interbreeding group is the essential unit in evolution. Action on intra-individual variation also occurs, but, again, can only eliminate, not originate, types of individuals or of individual reactions. (Simpson 1944, p. 31)

But the importance of variability is subordinate, in Simpson's view, to other determinants. Mere "segregation or selection of intragroup variability can give rise to new groups at a potentially rapid rate. This process depletes variability either by its definite elimination or by its transfer to the intergroup level, where it cannot provide materials for the origin of new groups. This process is important and typical in speciation and in lower levels of differentiation" (1944, p. 41). It is important to recall in reading this passage that, to Simpson, speciation is not typically the first stage in a long process leading to the emergence of new groups of progressively higher rank—the view that Dobzhansky (1937a, 1941) and Mayr (1942) were intent on establishing. Parceling of preexisting variation is important for the largely trivial pattern of racial differentiation within species and perhaps the origin of new species. But, as Simpson writes (1944, p. 42), "evolution on the basis of existing variability is a self-limiting process that cannot proceed beyond about the specific level."

Simpson lists several other concluding generalizations about variability, but only one is of significance to his argument and encountered again later in the book: "Rate of sustained evolution normally shows little or no correlation with variability. In more extreme cases of very slow or very rapid evolution this rate may tend to be negatively correlated with variability, but this has not been surely established" (p. 42). In this section, and later, Simpson equates low rates of evolutionary change with broad adaptations, concomitant lack of speciation, and broad anatomical variability, and also notes the converse: correlatively high rates, low variability, and narrow adaptations. Very much the same sort of correlative patterns have recently been cited (e.g., Eldredge 1979; Eldredge and Cracraft 1980, chapter 6; Vrba 1980) specifically as controlling factors in speciation rates. On the whole, to Simpson, some of the other determinants are more crucial than variability.

There is one further point to be made about Simpson's views on the evolutionary significance of variability: Simpson refers to Mather's paper (1941) to the effect that through polygenic combinations "such immediately available variability can be retained without sacrificing good adaptation" (Simpson 1944, p. 36). Simpson repeats the theme in his summary section on variability (p. 42): "Under the influence of strong selection changing in direction, balanced polygenic combinations can produce populations well adapted or specialized, without sacrificing variability available for rapid speciation." Simpson, like Dobzhansky, sees the dilemma posed by a too narrowly focused adaptation, which threatens to impede further evolution within a group.

Turning to mutation rate, Simpson simply tries to establish—with the aid of an elaborate scenario again involving equid evolution—that conventional estimates of mutation rates emanating from genetics laboratories are thoroughly consistent with observed rates of change in the fossil record. Simpson (p. 47) concludes that "most evolutionary sequences are consistent with moderate mutation rates and that their tempo and mode are not necessarily or primarily controlled by mutation rates." But, after so stating, Simpson lists "two theoretical situations to which this general inference would not apply" (p. 47). The first, the only one that figures later in his argumentation, is crucial.

Citing Wright, Simpson (p. 47) very subtly molds a proposition of population genetics to suit his larger purpose: quantum evolution, as developed later in his book, demands a mechanism for the loss of prior adaptation, followed by the development of preadaptation, as the precondition of the adoption of a radically different "adaptive peak" or occupation of a major, significantly different "adaptive zone." Both loss of adaptation and development of preadaptation in Simpson's theory of quantum evolution involve hypotheses concerning the role of mutation. Simpson cites Wright's (1940) view that, in very small populations, more than half the mutations occurring are likely to be deleterious—affording him a speculative mechanism for the loss of adaptation. The literature of population genetics was not exactly brimming with mechanisms to effect such loss of adaptation, and Simpson, much like Dobzhansky (for an entirely different reason) seems to have utilized Wright's statements (as well as Wright's entire landscape imagery) to help solve the problem. A further point needs emphasis here: as I shall soon develop, population size and natural selection emerge as the two most critically important determinants to Simpson—and in this "theoreticaly possible situation" involving high mutation rates we find the small population size that is so critical in the full-fledged theory of quantum evolution.

Next, Simpson considers the nature ("character") of mutations, and it is here that he attacks Goldschmidt (1940) and his concept of systemic mutations in a style that seems to have at least helped to set the tone of dismissal (bordering on ridicule) still heard today. Simpson's aim in this section is to establish that mutations of relatively minor effect are the main stuff of evolutionary novelty. Mutation supplies the raw material, enabling evolution ultimately to transcend the limits of the species to which simple variation in itself would confine the process. But why the strong attack on

Goldschmidt (including another of Simpson's several footnotes in which the word *fallacy* appears in conjunction with someone else's work)? There can be only one reason: the paleontological gaps central to Simpson's interests. Saltationists are the only ones who heretofore took such gaps as real and hence have a theory to explain them, or at any rate a theory other than the brief "the record is gappy and evolution sometimes quite rapid" typical of Darwinists, including Dobzhansky (1941, p. 343) and Mayr (1942, p. 297). Simpson wanted to treat some sorts of these gaps as real, which put him in immediate confrontation with the rival non-Darwinian notions of saltation. On his page 57, Simpson promises (and duly delivers in the next chapter) the argument against saltation from a paleontological point of view. But here, in his chapter 2, he had to attack saltation on the genetics front.

Simpson's attack on Goldschmidt leads to an amusing irony: Goldschmidt and Simpson were united and virtually alone among their peers in the belief that evolutionary changes within species seldom if ever led anywhere important in evolution—in the elaboration of important new anatomical structures, behavioral characteristics, or the development of taxa of higher categorical rank. To Goldschmidt (1940), macroevolution was all evolution above the species level (microevolution occurring within species), and the two processes were largely disconnected. Simpson (1944, p. 98) accepted the conventional definition of macroevolution as interspecific and intergeneric and saw an intergradation between speciation and macroevolution. He did see, though, a critical difference between macroevolution and what he called megaevolution (the origin of taxa of truly high rank; see below). Their views were not the same in detail, of course, but each did explicitly deny the importance of microevolution (adaptive differentiation of species via geographic variation) as the basis of what Goldschmidt called macroevolution and what Simpson termed megaevolution. The notion that macroevolution (including megaevolution) can be reduced to the principles of microevolution is, of course, the essence of the synthetic theory in its mature state; it is the position Simpson takes in 1953, and it is the stated position of Dobzhansky (1937a, 1941, 1951) and Mayr (1942). But in these early days Simpson's reductionism was not so simple an affair.

Following his discussion of mutation, Simpson turns to length of generations, the next on his list of evolutionary determinants. Earlier (his chapter 1, p. 20) he remarked (regarding rates of differentiation of North American mammals invading South America in Plio-Pleistocene times): "It is curious that this example gives no clear evidence for the more rapid evolution of animals with shorter generations." To Simpson and many of his contemporaries, evolution implied nearly inevitable change. For example, Simpson writes (1944, p. 149): "Organic change is so nearly universal that a state of 'evolutionary motion' is inherent in phyletic survival. It is probable that the continuous application of some sort of force, such as selection pressure, is necessary to maintain a state of rest and that the mere removal of restraint may be followed by acceleration." Such constant change, powered by selection working on a groundmass of variation in the neo-Darwinian paradigm, of course suggests that organisms with shorter generation times should logically evolve more quickly than those with longer generation times. They should show higher rates of morphological transformation. They don't. Simpson is candid in admitting that elephants have evolved far more quickly than opossums (1944, p. 63). He makes it clear that this supposed evolutionary determinant is inconsistent with patterns seen in the fossil record, but he does not inquire why—the whole matter, instead, is quietly dropped. It had to be; lack of substantiation of this factor of generation length is inconsistent with the view that evolution is essentially the transformation of phenotypic, and underlying genetic, characters within populations under the control of selection and genetic drift.

The way is then cleared for discussion of the two final—and, to Simpson, the most important—determinants of evolution: population size and natural selection. Absolute size of populations affects the rates of change of allelic frequencies within populations in a critical fashion. Not surprisingly, Simpson relies heavily on Wright here. I can do no better than quote Simpson's summary, which gives in a single paragraph the entire argument of the book, a precis of Simpson's theory of evolution as of 1944:

In summary, very large populations may differentiate rapidly, but their sustained evolution will be at moderate or slow rates and will be mainly adaptive. Populations of intermediate size provide the best conditions for sustained progressive and branching evolution, adaptive in its main lines, but accompanied by inadaptive fluctuations, especially in characters of little selective importance. Small populations will be virtually incapable of differentiation or branching and will often be dominated by random inadaptive trends and peculiarly liable to extinction, but will be capable of the most rapid evolution as long as this is not cut short by extinction. (Simpson 1944, pp. 70–71)

The three choices of population size listed here (large, medium, and small) correspond to some extent with Simpson's three classes of evolutionary rates (bradytely, horotely, and tachytely); with the notions about adaptation, already included in this paragraph, they combine to yield his three evolutionary modes: speciation, phyletic evolution, and quantum evolution, respectively. By page 70 of *Tempo and Mode*, then, the cat is already out of the bag, and the theme of the remainder of the book is thoroughly established.

Simpson closes chapter 2 of his book with a review of natural selection. To Simpson, as to all other architects of the synthesis, natural selection (differential reproduction on a generation-by-generation basis within populations) is the main, if not sole, deterministic factor shaping gene frequency changes in evolution. Simpson is at pains to do away with vitalistic,

Lamarckian, and preformist-preadaptation notions as the three basic alternatives he perceives to selection. (He returns to these alternatives in his chapter 5.) Discussing selection as a vector with both intensity and direction, Simpson argues strongly for the long-term great cumulative effects wrought by selection at relatively low intensities.

Simpson characterizes the direction component of the selection vector in the by now familiar patterns—centripetal, centrifugal, and linear—all with reference to a modal type. The imagery (see Simpson's fig. 11, reprinted here as fig. 3.2) fits nicely with Simpson's extension of Wright's (1932) metaphor of the adaptive landscape (see his fig. 12, reprinted here as fig. 3.3), which Simpson sees "at times more like a choppy sea than a static landscape" (Simpson 1944, p. 90). Eldredge and Cracraft (1980, p. 255) have already commented on the expanded nature of Simpson's use of Wright's metaphor, going far beyond the map of harmonious gene combinations that Wright saw (and Dobzhansky elaborated on) to provide a model for truly large-scale evolution, as Simpson's scenario of equid evolution that immediately follows his statement of Wright's metaphor makes clear (e.g., his fig. 13, reproduced here as fig. 3.4). Indeed, Simpson (1944, p. 93) goes beyond equids and applies the imagery to all of perissodactyl phylogenetic history.

So much, and far more, about selection is entrenched in the literature of the synthesis. It is therefore somewhat surprising to find, on the final two pages of Simpson's book, that selection has been omitted from a table

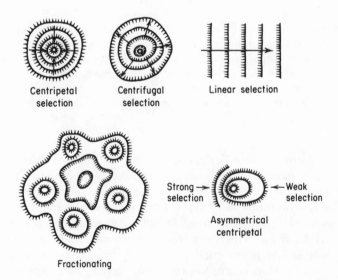

Figure 3.2   Simpson's fig. 11 (1944, p. 90): Selection landscapes. Contours analogous to those of topographic maps, with hachures placed on downhill side. Direction of selection is uphill, and intensity is proportional to slope.

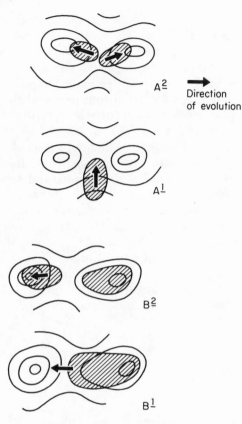

Figure 3.3   Simpson's fig. 12 (1944, p. 91): Two patterns of phyletic dichotomy, shown on selection contours. Shaded areas represent evolving populations. A. Dichotomy with population advancing and splitting to occupy two different adaptive peaks, both branches progressive. B. Dichotomy with marginal, preadaptive variants of ancestral population moving away to occupy adjacent adaptive peak, ancestral group conservative, continuing on same peak, descendant branch progressive.

(reproduced here as table 3.1) comparing ten characteristics for each of the three major modes of evolution recognized by Simpson. For Simpson, selection is the sine qua non means of achieving, or modifying, adaptation. In speciation—the mode most typical of the lowest levels of evolutionary phenomena—selection is involved in differentiation. But it is the middle range—phyletic transformation of one species into another, leading to the emergence of new genera—that evolution is held to be most adaptive in nature. Here selection reigns supreme. According to Simpson (1944, p. 203), ninety percent of the data of paleontology fall into this mode—progressive adaptive change within a phylum remaining within a single adaptive zone, changing under the linear (if shifting in direction and varying in intensity) guidance of natural selection. Claims such as these buttress an interpretation that Simpson shared his colleagues' feelings about the importance of selection, extrapolated purely and simply, directly from the *Drosophila* bottle to the data of paleontology strung out through true evolutionary time.

Yet the word *selection* hardly appears in Simpson's final chapter. The word *adaptation*, alone and with various prefixes, does appear, and it figures prominently. The key to understanding Simpson's views on selection in 1944 lies here: adaptation is of fundamental importance in the evolutionary process (e.g., "changes in adaptation, the most essential single phenomenon of modes of evolution"—p. 189). Its origin, change, and approach to perfection are governed by natural selection. To speak of changing adaptation (as, for instance, in the phyletic mode of his chapter 7), selection is completely implicit. But adaptation is so important to Simpson that, in quantum evolution, where rapid shifts from one adaptive

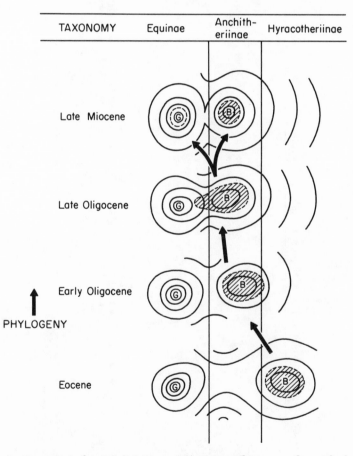

Figure 3.4   Simpson's fig. 13 (1944, p. 92): Major features of equid phylogeny and taxonomy represented as the movement of populations on a dynamic selection landscape.

Table 3.1  Simpson's characteristics of the main modes of evolution

| Mode | Typical taxonomic level | Relation to adaptive grid | Adaptive type | Direction | Typical pattern | Stability | Variability | Typical morphological changes | Typical population involved | Usual rate distribution |
|---|---|---|---|---|---|---|---|---|---|---|
| Speciation | Low: subspecies, species, genera, etc. | Subzonal | Local adaptation and random segregation | Shifting, often essentially reversible | Multiple branching and anastomosis | Series of temporary equilibria, with great flexibility in minor adjustments | May be temporarily depleted and periodically restored | Minor intensity; color, size, proportions, etc. | Usually moderate with imperfectly isolated subdivisions | Erratic or comparable to horotelic rates |
| Phyletic evolution | Middle: genera, subfamilies, families, etc. | Zonal | Postadaptation and secular adaptation (little inadaptive or random change) | Commonly linear as a broad average, or following a long shifting path | Trend with long-range modal shifts among bundles of multiple isolated strands, often forked | Whole system shifting in essentially continuous equilibrium | Nearly constant in level; most new variants eliminated | Similar to speciation, but cumulatively greater in intensity; also polyisomerisms, anisomerisms, etc. | Typically large isolated units, with speciation proceeding simultaneously within units | Bradytelic and horotelic |
| Quantum evolution | High: families, suborders, orders, etc. | Interzonal | Preadaptation (often preceded by inadaptive change) | More rigidly linear but relatively short in time | Sudden sharp shift from one position to another | Radical or relative instability with the system shifting toward an equilibrium not yet reached | May fluctuate greatly; new variants often rapidly fixed | Pronounced or radical changes in mechanical and physiological systems | Commonly small, wholly isolated units | Tachytelic |

Source: Simpson 1944, pp. 216–17.

zone (or peak) to another are envisioned, the first step is a loss of ancestral adaptation—something impossible for a good Darwinian to attribute to natural selection. The initial and middle stages of this favorite idea of Simpson's (i.e., quantum evolution, the idea for which the book is primarily remembered) specifically denies any role for natural selection at all. Simpson says, in effect, that selection is extremely important in evolution, but he assigns it essentially a mop-up role in the very process that creates higher taxa from old ones; loss of adaptation and development of preadaptation come first and take precedence. In chapter 2, Simpson does claim a role for selection in preadaptation, but the essentially fortuitous nature of accumulated mutations (p. 36) or structures (p. 79) suddenly found to be in an environment where they are useful (and thus can come under selective control) remains clear in the book. In 1953, Simpson dropped the "inadaptive phase" of quantum evolution, and the way was paved for a far more orthodox (or what we have since come to view as orthodox in the mature synthesis) invocation of the complete dominance of selection as the moving force behind all evolutionary change, including the rapid (tachytelic) bursts of quantum evolution.

In his third chapter, Simpson states the basic problem of the book—the gaps between taxa in the fossil record—evaluating their nature and outlining his theory for their explanation. The chapter is entitled "Micro-Evolution, Macro-Evolution, and Mega-Evolution." In *The Major Features of Evolution*, the expanded successor to *Tempo and Mode*, Simpson acknowledges that in 1944 he

expressed the opinions that microevolution and macroevolution are not qualitatively distinct, that there is no more reason from this point of view to draw a line at the species level than at the genus, family, etc., levels, and that macroevolution as applied to species and genera does not look very "macro" to a student of the major features of evolution. (Simpson 1953, p. 339)

Though Simpson expressly denies that he meant, in 1944, that mega-evolution is different in kind from micro- and macroevolution, his own words are quite clear:

If the term "macro-evolution" is applied to the rise of taxonomic groups that are at or near the minimum level of genetic discontinuity (species and genera), the large-scale evolution studied by the paleontologist might be called "mega-evolution" (a hybrid word, but so is "macro-evolution"). The assumption, as in Goldschmidt's work, that mega-evolution and macro-evolution are the same in all respects is no more justified than the assumption, so violently attacked by Goldschmidt and others, that micro-evolution and macro-evolution differ only in degree. *As will be shown, the paleontologist has more reason to believe in a qualitative distinction between macro-evolution and mega-evolution than in one between micro-evolution and macro-evolution.* (Simpson 1944, p. 98; italics added)

Accepting the conventional definition of macroevolution, Simpson invented megaevolution to cover the case of the origin of taxa of higher categorical ranks. The rest of the chapter is a detailed argument that (1) minor discontinuities in the fossil record (i.e., between species and genera, hence macroevolution in his usage of 1944) are almost always artifacts of a poor fossil record, but that (2) major discontinuities in the fossil record (between families and taxa ranked even higher, hence megaevolution) are real phenomena requiring explanation. To Simpson in 1944, megaevolution was indeed a new sort of evolution.

Microevolution is characterized but not treated at length either in this chapter or in the book as a whole. Simpson is explicit and consistent throughout. The trivial changes that are the stuff of microevolution—those patterns that occupy zoologists, botanists, and geneticists—are largely irrelevant to understanding the large-scale phenomena seen by paleontologists, though all these phenomena result from the same class of determinants arranged in various combinations and strengths (Simpson 1944, p. 124). For example, in his preface Simpson wrote:

On the other hand, experimental biology in general and genetics in particular have the grave defect that they cannot reproduce the vast and complex horizontal extent of the natural environment and, particularly, the immense span of time in which population changes really occur. They may reveal what happens to a hundred rats in the course of ten years under fixed and simple conditions, but not what happened to a billion rats in the course of ten million years under the fluctuating conditions of earth history. Obviously, the latter problem is much more important. (Simpson 1944, p. xvii)

That macroevolution is the basically linear, progressive, adaptive modification of lineages occupies more of Simpson's time. In a lengthy footnote (his footnote 9, p. 58) in chapter 2, on the subject of Goldschmidt's and Schindewolf's saltationist views, Simpson starts out: "To those who have done much work on good phyletic series of fossils it will hardly seem necessary to make such an obvious statement as that good species and genera frequently arise in this gradual way, whether or not they always do."

In his third chapter, Simpson says that species, genera, and "at least the next higher categories" (p. 99) often arise by a gradual, continuous, vertical process of phyletic evolution. And later, Simpson asserts (1944, p. 203) that fully nine-tenths of the data of paleontology are to be interpreted in terms of gradual, adaptive change in the phyletic mode—wholesale transformation of entire lineages.[5] But in spite of these appeals to paleontological experience and the frequent assertions, such as these just cited, no actual, ironclad cases of phyletic transformation are actually given, though examples are presented anecdotally. The main point here is that, despite its claimed frequency and his assertion of its overwhelming importance in the evolutionary process, Simpson spends rather little time on adaptive phyletic change per se in his book.

In chapter 3, Simpson disposes of minor discontinuities (i.e., those among species and genera) by arguing that whenever the record is good the expected patterns of gradual transition between ancestor and descendant emerge. It is, again, a consistency argument at base. In his characteristically excellent summation of the argument (p. 105), he says: "On these levels everything is consistent with the postulate that we are sampling what were once continuous sequences." But (p. 105) "the levels to which these conclusions apply without modification are approximately those discussed as macro-evolution (under that or an equivalent term) by neozoologists and biologists. On still higher levels, those of what is here called 'mega-evolution,' the inferences might also apply, but caution is enjoined, because here essentially continuous transitional sequences are not merely rare, but are virtually absent." In other words, the pattern appears to be qualitatively different—in which case, the explanation of it must also be different. Instead of progressive adaptive change, we need another model.

After characterizing the nature of the major systematic gaps in the fossil record (including his paleontological falsification of saltationism on p. 115, which rests essentially on the observation that intermediate forms between higher taxa, such as *Archaeopteryx*, are occasionally found), Simpson gives us the new explanation: if saltation is not the answer, circumstances must conspire to produce the gaps. On pages 117 and 118, Simpson argues that small population sizes, rapid evolutionary rates, small organismic body sizes, and factors of geographic occurrence (including narrow localization in regions undergoing erosion rather than deposition) would conspire to create the gaps in the record.

His analysis—rather similar to the brief discussions already cited in Dobzhansky (1941, p. 343) and Mayr (1942, p. 297), not to mention Darwin's (1859, p. 342) summation of the geological causes of the gaps in the fossil record—is merely the beginning of Simpson's argument. That unusually high rates must be involved can be seen by the extrapolation of normal rates leading to absurdly unacceptable results: based on rates of change between Cretaceous and Recent opossums, the reptilian transition "to an opossum can hardly have taken less than 600,000,000 years; it probably took several times that long—in short it must have occurred in the pre-Cambrian, which is certainly absurd" (Simpson 1944, p. 119). Precisely these sorts of arguments persist to the present (e.g., Stanley 1975a) against the pervasiveness of gradualism as the basic mode of origin of life's diversity.

What is the solution? Again, I can do no better than cite Simpson's own summary:

The theory here developed is that mega-evolution normally occurs among small populations that become preadaptive and evolve continuously (without saltation, but at exceptionally rapid rates) to radically different ecological positions.

The typical pattern involved is probably this: A large population is fragmented into numerous small isolated lines of descent. Within these, inadaptive differentiation and random fixation of mutations occur. Among many such inadaptive lines one or a few are preadaptive, i.e., some of their characters tend to fit them for available ecological stations quite different from those occupied by their immediate ancestors. Such groups are subjected to strong selection pressure and evolve rapidly in the further direction of adaptation to the new status. The very few lines that successfully achieve this perfected adaptation then become abundant and expand widely, at the same time becoming differentiated and specialized on lower levels within the broad new ecological zone. (Simpson 1944, p. 123)

So here again is quantum evolution, not called by that name in Simpson's chapter 3, but said (p. 124) to be most typical of "mega-evolution," though also occurring in macro- and microevolution. Note too the use of Wright's picture of a large population fragmented "into numerous small isolated lines of descent"—a condition that Wright (1931, 1932) also saw as conducive to "inadaptive differentiation" and "random fixation of mutations."

Simpson's fourth chapter is entitled "Low-Rate and High-Rate Lines," though it deals almost exclusively with low-rate lines. Such a preoccupation would seem to be a departure from the theme of quantum evolution, with its invocation of "exceptionally rapid" rates, but it is actually the most critical evidence adduced in the book for the very existence of quantum evolution. By its very nature, megaevolution (or, as it is called later in the book, quantum evolution) leaves no direct trace in the fossil record; its existence is wholly inferential. And Simpson's claim that there is a qualitative difference between megaevolution on the one hand and micro- and macroevolution on the other, is merely a suggestion based on the position that major gaps are real biological phenomena requiring explanation, but minor gaps are merely deficiencies in the data. Simpson clearly felt the need for some substantiating line of evidence for his argument.

The very data for extremely fast rates of evolution are missing—for the fast rates produce the gaps. Simpson therefore shifts to extremely low rates of evolution—the mirror case—and attempts to demonstrate that there exists a class of slow evolutionary rates with a wholly different distribution from that of the middle and upper rate classes. In other words, extremely slow evolution is not just the lower tail of a normal distribution of evolutionary rates. It is, rather, a qualitatively distinct group of rates, with the attendant implication that each class of rates requires its own intrinsic combination and intensity of evolutionary determinants (genetic factors). The demonstration of a distinct class of low rates (bradytely) by implication and analogy established the class Simpson was mainly interested in: tachytely, the class of exceptionally fast rates. Stanley (1984) provides a recent critical appraisal of Simpson's contention that bradytely, horotely, and tachytely actually represent three separate distributions of evolutionary rates.

Simpson went back to the survivorship curves developed in his opening chapter to establish these three different rate distributions. Returning to the pelecypod data, he writes (1944, p. 133): "The conclusion is inescapable that the present pelecypod fauna is the resultant of at least two decidedly different rate distributions." One may, of course, take issue with some of the assumptions of this entire argument (e.g., survivorship of genera is assumed to be the inverse of evolutionary rates, and expected rates—against which the realized rates of the data are compared—are assumed to be normally distributed). Debate continues to the present day on the suitability and significance of survivorship analysis of paleontological data.

In his characterization of bradytely, Simpson envisions a concatenation of variables whose inverse automatically leads to a characterization of tachytely. For instance, on his page 138, we read that large breeding population size seems to be characteristic of bradytelic groups. But it is not until we reach the summary that tachytely itself is mentioned:

Tachytelic lines either become extinct or usher in new major adaptive grades in which the phyla become on the whole horotelic, but often one line or a minority of lines are bradytelic. (Simpson 1944, p. 147)

Tachytely exists as a discrete class of high rates, with such correlates as small population size, mainly as an analogy with the analyzed case presented to establish bradytely as a discrete class of (exceptionally low) rates.

I shall pass lightly over Simpson's chapter 5 ("Inertia, Trend, and Momentum"). It falls into the general scheme of paving the way for Simpson's presentation of the three modes of evolution, but only in a negative sense. In 1944, it was still necessary for a paleontologist confronting evolutionary theory to deal with competing *paleontological* theories (as opposed to saltation, which has had adherents from several disciplines within biology). Such theories include notions pertaining to sustained directionality and the alleged tendency for some lineages to pursue modification of structure to the point where the feature becomes harmful, contributing to the demise of the lineage. That these issues are no longer the red-hot topics they once were is largely the result of Simpson's careful dissection of them here and in sundry later publications. A major aspect of bringing paleontology within the bounds of evolutionary theory in the 1940s was the excision of various metaphysical notions which, by their very nature, had no place in the formulation of scientific theory. The consistency format once again reigns: none of the apparent patterns that seemed to cry out for special explanation seems to Simpson to necessitate explanations in any terms other than those of the neo-Darwinian paradigm.

But Simpson's fifth chapter, in eliminating the mystical, clearly leaves selection as the prime evolutionary determinant. Mutations impose constraints, but mutation in itself cannot produce major patterns; it is selection that underlies the major trends in evolution, as Simpson's chapter summary reiterates. Here orthoselection (linear, long-term directional natural selection) comes to replace orthogenesis as the paleontologist's explanation for the production of patterns of "rectilinear" evolution—persistent, long-term directional trends in the transformation of anatomical properties.

In chapter 6, Simpson returns to the problem of adaptation as the final prerequisite to his characterization of the three main modes of evolution in the seventh and concluding chapter of *Tempo and Mode in Evolution.* Although he begins (p. 180) by pronouncing it an "extreme dictum that all evolution is primarily adaptive," he also says a few pages later (p. 189) that changes in adaptation constitute "the most essential single phenomenon of modes of evolution."

Simpson presents a clearly hierarchical classification of adaptation, involving "individual" and "group" levels. Curiously, though he claims his classification has "practical significance for studying the phenomena of adaptation" (p. 182), he soon launches into a discussion of preadaptation, having no further use for his explicitly hierarchical eightfold classification of adaptation.

Preadaptation already figured prominently in Simpson's second chapter on evolutionary determinants. The concept is critical to most neo-Darwinian attempts to circumvent the problem posed by ancestral structures that cannot function in the same manner as the new, fully developed descendant structures. The general answer (Simpson 1944, p. 186) stems from the notion of "prospective functions": the newly arising structure serves some other purpose and is thus under the control of selection until it is formed to the point where it can serve a second function, whereupon selection for the second function takes over to perfect the adaptation directly (see Gould and Vrba 1982 for a recent discussion and classification of such phenomena). This is how selection is alleged to be involved in preadaptation, though Simpson (1944, p. 186) acknowledges its random aspect. Adaptation—its maintenance and modification, its loss and reemergence in modified form—figures heavily in Simpson's three evolutionary modes (speciation and phyletic and quantum evolution); pre- and postadaptation are cornerstones in his theory of how such changes are effected.

In his chapter 2, discussing the nature and role of natural selection as a determinant in the evolutionary process, Simpson makes heavy use of the adaptive landscape imagery of Wright. Now, in chapter 6, he introduces the concept of the adaptive grid, upon which are plotted adaptive zones and subzones. Not clearly defined, adaptive zones emerge from the "consideration of the environment as composed of a finite and more or less clearly delimited set of zones or areas (1944, p. 189). Eldredge and Cracraft (1980, pp. 261 ff.) conclude that the two sets of imagery are essentially equivalent (as indicated by Simpson's use of the same equid examples with both sets of imagery). The adaptive grid introduces time as one dimension, whereas the landscape depicts the adaptive scene at only one point in time. Simpson does use the grid to discuss the pattern of steplike evolution ("Stufenreihe")—structural stages or grades rather than directly intergrading sequential phyletic series. Stufenreihe are the sort of pattern that many paleontologists (e.g., Vrba 1980; Stanley 1979; Eldredge and Cracraft 1980) see as commonly underlying trends. But the major significance of adaptive zones in Simpson's theory becomes clear in his chapter 7, "Modes of Evolution."

The seventh and final chapter of *Tempo and Mode* is a concise and effective summary of the main argument of the book. Three modes—supposedly inferred from consideration of evolutionary determinants and tempos but in reality discussed throughout the book—are delineated: speciation, phyletic evolution, and quantum evolution. Each is typical of but not restricted to particular levels of the taxonomic hierarchy: speciation is the process of raciation leading to the emergence of new species; phyletic evolution is the origin of new species per se and their further transformation into genera; quantum evolution is the rapid origin of families and taxa of even higher categorical rank.

Eldredge and Cracraft (1980, p. 261) give this brief characterization of Simpson's three modes (see also Simpson's illustration, reproduced here as fig. 3.5):

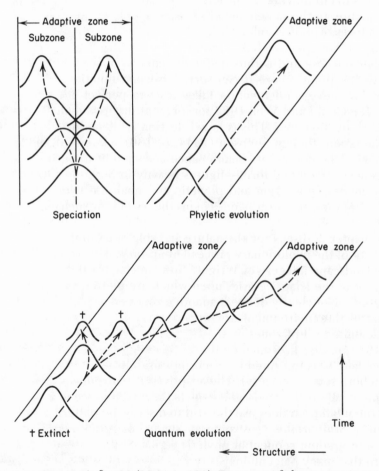

Figure 3.5   Simpson's fig. 31 (1944, p. 198): Diagrams of characteristic examples of the three major modes of evolution. The broken lines represent phylogeny, and the frequency curves represent the populations in successive stages.

1. Speciation: differentiation, usually a within-species phenomenon. In terms of the adaptive grid, it involves either differentiation of a population into subzones of a single zone, or the elaboration of new adaptations, allowing later invasions into new subzones.

2. Phyletic evolution: the "sustained, directional (but not necessarily rectilinear) shift of the average characters of populations," a mode "typically related to middle taxonomic levels, usually genera, subfamilies, and families. In relation to the adaptive grid, phyletic evolution is usually or most clearly seen as a progression of single or multiple lines within the confines of one rather broad zone" (Simpson 1944, p. 203).

3. Quantum evolution: "the relatively rapid shift of a biotic population in disequilibrium to an equilibrium distinctly unlike an ancestral condition" (Simpson 1944, p. 206). Like other modes, it can give rise to taxa of any rank, but Simpson proposed quantum evolution as "the dominant and most essential process in the origin of taxonomic units of relatively high rank, such as families, orders, and classes." In terms of the adaptive grid, quantum evolution pertains to interzonal shifts. (Eldredge and Cracraft 1980, p. 261)

Speciation, according to Simpson, is difficult if not impossible to study with fossils; it proceeds generally by subdivision of adaptive zones. Phyletic evolution, that ninety percent of paleontological phenomena, proceeds typically by modification of adaptation (controlled by selection) within a changing adaptive zone. Quantum evolution, suggested inferentially by the large gaps between higher taxa and the nearly simultaneous appearance of both ancestral groups and their descendants in the fossil record, involves loss of adaptation, preadaptation, and postadaptation as the initial adaptive equilibrium in one zone is lost, only to be regained in substantially different form within another adaptive zone (or on another peak). Population size is typically small; mutations can become fixed in the inadaptive phase, facilitating preadaptation.

New species might arise via quantum evolution, as may genera and families; conversely, taxa of higher rank may emerge via the process of phyletic evolution, though Simpson says that this cannot be the usual case in view of the dearth of examples. The large gaps are real; the smaller ones, for the most part, are only apparent.

This, the crowning statement of *Tempo and Mode in Evolution*, to which the argument was leading throughout its initial six chapters, is so consistent with Simpson's foregoing discussions as to constitute more of a summary than a conclusion. Speciation, phyletic evolution, and quantum evolution are identical for all practical purposes with micro-, macro-, and megaevolution in his chapter 3. (The threefold classification of types of evolution is not used in chapter 7, leading one to wonder if the exigencies of writing a book in wartime interfered in this respect with what is otherwise a singularly coherent and consistent presentation.) The only notion added after chapter 3 that contributes to Simpson's conclusion that large gaps are different in more than mere order of magnitude from the smaller gaps (i.e., his micro- and macro- vs. megaevolution dichotomy) was his demonstration that survivorship patterns of pelecypod genera indicate the existence of two qualitatively distinct rate classes. Simpson regarded the existence of three qualitatively different classes of rates as crucial support of his notion of qualitatively different modes—the three combinations of determinants each producing the three quasi-discrete modes of evolution.

Simpson summarizes his three modes—and effectively his entire book—in a table (1944, pp. 216–217, reproduced earlier here as table 3.1) and accompanied by a typical caveat (p. 217): "In each case it is not implied that the stated condition for a given mode is the only possibility within that mode or that it is confined to that mode, but only that it is a common, favorable, or distinctive condition."

Thus, Simpson's book-length argument hardly constitutes a bland, across-the-board reductionism. Characterizing normal speciation (by adaptive differentiation) as an essentially trivial process, and not the requisite first step in the evolution of taxa of higher rank, is a denial of a fundamental postulate of the synthesis as it exists today—and of the positions of Mayr (1942) and Dobzhansky (1937a, 1941). *Tempo and Mode* is, instead, a clever argument designed to show that paleontologists are uniquely suited to study the major phenomena of evolution. Simpson had a hierarchy of dimension of evolutionary pattern in mind here, as he did in 1953, for that is what he means by "the major features of evolution." The underlying causes of these phenomena can indeed be reduced to the first principles of the neo-Darwinian paradigm, but not in the fashion already conventionally supposed by Simpson's contemporaries. No indeed—the major, most important pattern of the fossil record requires its own special configuration of theoretical elements. Selection and adaptation, the twin ruling concepts of the synthesis, as important as they are in Simpson's view of the evolutionary process, take on a surprising guise in the most important mode of Simpson's tripartite view of the evolutionary process. For truly major adaptive change, adaptation itself must be lost before it can be regained in radically different form. Selection itself is incapable of leading populations (most of which suffer extinction) through the metaphorical valley of the shadow of death to a new realizable adaptive peak. *Tempo and Mode* did bring paleontology under the umbrella of the synthesis. But at the same time Simpson held paleontology aloof; its most important pattern could not be explained by the most direct and simplistic extrapolation of natural selection working constantly, gradually, and progressively to modify adaptation—the very view that later emerged as the paramount theme of the synthesis.

## A NOTE ON *THE MAJOR FEATURES OF EVOLUTION*

Simpson wrote in the preface of *The Major Features of Evolution* (1953) that he felt that *Tempo and Mode in Evolution* had served its purpose and, by the 1950s, should have been allowed to die quietly. When pressed for

a revision, he produced instead the book for which he is most famous—
*Major Features*. Much of the content of *Tempo and Mode* remains in its
successor, including large chunks of unmodified text and many of the original illustrations. And it is widely appreciated that in 1953 Simpson tem-
pered his "formerly more extreme views" (1953, footnote 6, p. 392) on
quantum evolution. Specifically he eliminates the inadaptive phase and
assigns selection a greater role throughout—so much so that quantum
evolution is toned down to being a "special, more or less extreme and
limiting case of phyletic evolution" (Simpson 1953, p. 389). Still seen as
a process of rapid change between adaptive zones, it is no longer construed as the sort of quasi-discrete mode as it was in 1944.

Concomitantly, as I have already noted, Simpson drops the terms
*micro-*, *macro-*, and *megaevolution*, to the point of denying that he meant
earlier to suggest an element of qualitative difference among them. Even
his views on preadaptation are modified, to bring the phenomenon more
under the direct control of selection. Here (1953, p. 270) he argues that
even the very earliest stages of trends confer some adaptive advantage.

Only the section devoted to the demonstration of the existence of qualitatively different, discrete classes of evolutionary rates is retained—
rewritten, but using the same pelecypod data and the original illustrations. But its purpose is gone, and Simpson makes little further use of it;
it no longer makes any real difference, insofar as modes of evolution are
concerned, whether the rate classes actually exist or whether there is in
fact a single, normal distribution for evolutionary rates (though Simpson
retains an interest in the phenomenon—especially bradytely—per se).

If the central argument of *Tempo and Mode* is gone from *Major Features*,
the structure of the latter book, largely inherited from its forerunner,
must be interpreted differently as well. *Major Features*, as the title itself
suggests, reads more like a compendium of evolutionary phenomena—
diverse patterns and their interpretations—than the closely knit, highly
structured single argument that comprised its predecessor.

In the end, the fundamental difference between the two works is Simpson's tempering of the very point he most strongly argued as his central
thesis of *Tempo and Mode*—that paleontological data can be reduced to
the neo-Darwinian paradigm but not in the fashion suggested by geneticists and other students of the Recent biota. In 1953, Simpson still takes
swipes at what might appear to be the trivial problems of evolutionary
transformation studied by geneticists, but the major point is conceded,
and, with *The Major Features of Evolution*, the synthesis between paleontology and genetics is complete; adaptation via selection, as a simple
extrapolation and at the rates studied in the laboratory, leads inexorably
to the phyletic change that paleontologists document with their fossils. In
1944, Simpson had by no means been willing to give the game away so
thoroughly. He had tried hard to retain a measure of his discipline's idi-
osyncrasy of claiming some segment of evolutionary pattern—and, therefore, of evolutionary process—as its own special province. It is the spirit

of *Tempo and Mode in Evolution*, more than that of *The Major Features of Evolution*, that inspires paleontological interest in evolutionary theory in the 1980s.

We have now examined in detail each of the four early works by the three major figures in the actual writing of the synthesis. After a brief summary analysis comparing their views, we will be in a position to ask if there ever was an evolutionary synthesis at all, and, if so, what its central tenets actually are. We can then take a brief look at what has been added to the structural frame outlined in these works by these and all the other evolutionary biologists writing since the 1940s.

## NOTES

1. Once again, an explicit distinction between taxa, on the one hand, and the variously ranked categories of the Linnaean hierarchy on the other, was not yet in vogue in 1942. Mayr accordingly uses the word "category" where he would later have written "taxa" in many places in the book, as he himself points out in the preface (p. x) to the previously reprinted edition (1964) of this work (though there Mayr claims there is some confusion between taxa and categories in chapter 10, where in fact the confusion pervades the entire book). Simpson's (1963) paper clarified the situation considerably.

2. Dobzhansky (see chapter 2 herein), Mayr, and Simpson all duly noted the superficial resemblance—yet important differences—between the saltational views of geneticist Richard Goldschmidt and the views of paleontologist Otto Schindewolf.

3. I say 1944, the book's publication date, fully aware that Simpson started the manuscript in 1938 and completed it in 1942 (Simpson 1953, p. ix).

4. Evolutionary rates here mean rates of morphological change. Simpson's brief discussion of taxonomic rates in this chapter, greatly augmented in 1953, is explicitly geared to estimate the *real* rates: morphologic (thus underlying genetic) rates of evolutionary change (see Eldredge 1982b for fuller discussion). In line with my discussion of Simpson's view of the nature of taxa, he introduces taxonomic rates in the following way (1944, p. 16): "Insofar as it seeks to divide phyla into generic and specific stages, representing roughly equivalent amounts of total morphological change, the taxonomic system is a rich source of such data." As we shall see, Simpson's basic theme in *Tempo and Mode* hinges on these very taxonomic rates, as developed later in his book (his chapter 4).

5. Simpson's fig. 33 (1944, p. 204) shows "three patterns of phyletic evolution"—including "a population early well-adapted to a stable zone with little subsequent change." Inclusion of such stasis under the "phyletic mode," of course, would force most paleontologists (such as myself) impressed with the prevalence of stasis to agree with Simpson that at least ninety percent of the data of paleontology are of the phyletic mode. That Simpson really means "phyletic transformation" (i.e., most certainly not stasis) by his expression "phyletic mode" is abundantly clear from his discussion, including, of course, excerpts cited herein.

# 4

# The Structure and Content
# of the Modern Synthesis

"Many evolutionists would be surprised to see identified as the two 'fundamental claims' of the modern synthesis those listed by Gould [i.e., "extrapolationism (gradual allelic substitution as a model for all evolutionary change) and nearly exclusive reliance on selection leading to adaptation"] and most would not agree that the modern synthesis has broken down. The impression that a 'straw man' has been erected is confirmed when one discovers that the proposed new 'themes' are part and parcel of the modern synthesis." (Stebbins and Ayala 1981, p. 967)

When taken together, the four books of Dobzhansky, Mayr, and Simpson, written as they were so closely together in time (and space—Columbia University and the American Museum of Natural History are within forty blocks of each other in New York City) reveal a relatively minor amount of disarray, a slight lack of cohesion in the early stages of the synthesis. That some of these discrepancies were later removed—most notably through a more universal acceptance of the dominant role of natural selection in effecting adaptive change (Gould 1980b)—is important, if only because it established more of a semblance of agreement and consensus. The acceptance that Mayr (1982, pp. 568–69) reports among nearly all participants at the Princeton conference held in 1947 seems real enough; by the late 1940s the final, polished version of the synthesis apparently had begun to emerge.

But we must ask if there were any important additions to evolutionary theory after these four books appeared. Changes in emphasis—for example, on selection, but also in such issues as Mayr's later (especially 1963) views on the role that species play in evolution—certainly did occur. And, of course, beyond the conceptual lies the straightforward discovery of new phenomena, such as the myriad wonders of the molecular biology of the gene, begun in earnest in the early 1950s and still being announced daily. What concerns me here is more the structure of evolutionary theory than its precise content. Have either new ideas or new data since the publication of these four books materially modified the way we think about evolution? The answer, for the most part, is no; the theory presented in the better recent college textbooks (e.g., Dobzhansky et al. 1977; Futuyma 1979) is substantially the same as the amalgam that arose from the four books analyzed here, with the rough edges sanded and recent discoveries—nearly all concerning the molecular structure of the gene—duly incorporated.

But there were some particularly important innovations and shifts of emphasis within the purview of the synthesis. And there have been several developments that, it seems to me, virtually demand a restructuring of evolutionary thought, pointing to the need to adopt a hierarchically structured evolutionary theory. First in order is a simple recapitulation of the core structure of the theory in Dobzhansky, Mayr, and Simpson—a composite theory that, despite its contradictions and dilemmas both within and among books, nonetheless comes close to the structure of the synthesis as we know it today. I will then examine what I think have been the (relatively few) significant shifts in emphasis and content that have affected the structure of evolutionary thought to some degree. In chapter 5, I will take up some of the more troublesome ideas and observations that, when pursued thoroughly, seem to call for a more far-reaching restructuring of evolutionary theory.

## THE WORLD'S FURNITURE: WHAT EVOLVES?

What is it that evolutionary theory seeks to explain? For the remainder of this book I will deal with three major categories of concern to evolutionary theorists: biological entities (e.g., genes, species, phyla, communities), biological processes (e.g., natural selection, mutation, species selection, interspecific competition), and classes of historical biological events or "patterns" (e.g., linear trends, allopatric speciation). (Many patterns confusingly share the same name as their underlying causative process—such as mutation, speciation, etc.) I believe it is critical to keep these three categories of items of evolutionary interest distinct so that we can better understand their interrelationships. The items admitted into each category and the interrelationships among the items within each category determine both the structure and the content of evolutionary theory.

Bunge (e.g., 1977) has called the items of the first category—the physical things out there (biological and nonbiological)—the world's "furniture." I will have far more to say about the ontological status of various of such biological entities in later chapters. Here we need only note that there is a conventional list of entities, of biological furniture, that forms the hub, the very focus, of all evolutionary writing—indeed, of all biological research at least since Aristotle. A modern version of this list minimally includes codons, genes, chromosomes, organisms (with their cellular organelles, cells, tissues, organs, and organ systems), populations, demes, species, monophyletic taxa, communities, ecosystems, and regional biotas.

All theorists think that some version of this list of entities has something to do with evolution. Indeed, evolution to all of us is a matter of what happens to some or all of these entities through time. It is precisely because all these elements *are* relevant to evolution, yet the theory adduced to form the synthesis explicitly addresses but a few of the entities (or at least pays far more attention to some than to others), that the synthesis is found to be incomplete—not "wrong," but incomplete.

Dobzhansky (1937a, 1941) took a distinctly atomistic view of the gene, seeing genes strung in linear arrays along chromosomes, but otherwise (of necessity) he treated the problem of the biochemical anatomy of genes as a black box. That advances in molecular biology contribute to the need for a formal expansion of evolutionary theory is an exigency we can hardly hold against the early architects of the synthesis. Mayr (1942), on the other hand, appropriately treated *genetics* as a black box, referring his readers to Dobzhansky and explicitly adopting the particularist view of heredity. It was only later that Mayr came to focus on genetics per se, labeling the older work on single loci of Fisher, Wright, and Haldane as "bean-bag genetics" (see Kottler 1983, p. 869) and focusing himself on the importance of "coadapted gene complexes" and "genetic revolutions" in the 1950s. Simpson, too, explicitly attempting a synthesis of genetics with the data of paleontology, concentrated on the processes of genetics (mutation, selection, and the like) and took the standard view on the nature and existence of the basic elements—genes and chromosomes—as given. All this amounts to, of course, is that Dobzhansky adopted the contemporary consensus of what he called "physiological genetics" bearing on the nature of genes and chromosomes and explicitly integrated these elements with the biology of populations, the origin of new species, and other evolutionary phenomena. This itself is no surprise, nor is it a surprise that Mayr and Simpson accepted the current principles of physiological and population genetics axiomatically.

Organisms carry the genes, and it is the phenotypes of organisms, varying among one another, that are "selected." If it is trivially true, it still must be formally stated that all three authors accept the existence of organisms and see their role in evolution as the external expression of the genes they carry, as the transmitters of those genes, as the bearers of adaptations in their phenotypes (in the broadest sense of the term, embracing morphology, physiology, and behavior), and as the expressers of variation within populations. Organisms are thus central, even pivotal to their evolutionary views—though Dobzhansky deliberately eschewed focusing on the phenotype per se (while acknowledging that what was known of physiological genetics hinged on evaluations of the phenotype), while Mayr and Simpson each spent relatively more time on adaptation, thus stressing the phenotype.

But, in another sense, evolutionists in general—not just Dobzhansky, Mayr, and Simpson—tend to see organisms as bystanders in the evolutionary process. Mutation modifies genes; selection and drift modify frequencies of alleles within populations. Organisms per se do not evolve, though they do participate in processes that lead to genetic change.

Organisms, instead, have ontogenies; they differentiate, develop, and otherwise become transformed from fertilization (in the case of sexual species) until death. The fact that a number of "evolutionary" theories are expressly theories of the transformation of ontogenetic pathways merely verifies the necessity for such change if the phenotype is to be transformed in the evolutionary process. That the ontogeny of organisms is left wholly out of the discussions of each of the four books dealt with here is interesting; developmental biology in a formal sense remains a black box in the structure of evolutionary thought to the present day. Organisms, in this particular sense, seem to be less important than genes on the one hand and the genetics of populations on the other. But the point can be pushed too far; selection, the main deterministic agent of change of the genetics of populations, works on the differences of the phenotypes among organisms within populations. Thus, there is no discernible difference among Dobzhansky, Mayr, and Simpson in how they treat the nature and role of organisms in evolution. That the subject sounds trivial is only because we have all become so accustomed to precisely this description of the nature and function of organisms in evolution. Finally, organisms also have ecological roles as prime loci of the biological conversion of matter and energy in nature—what might be termed "economic adaptations." This aspect of organisms is of course acknowledged but little discussed in these four books.

What, then, of populations? Dobzhansky (1937a) remarked that Wright's "colonies"—semi-isolated segments of sexual species—were a common phenomenon in nature. The existence of such groups (later to be termed "demes" in Wright's theories) was important to Dobzhansky because so much of his own theory of the dynamics of evolutionary change hinged on Wright's views, and the differential fates of such clusters within a species did much to reduce variation (whether randomly or by selection), whereas the mere fact of existence of such clusters allowed more peaks in the adaptive system within a species to be explored. It was convenient, in other words, for Dobzhansky to emphasize that species frequently are fragmented into numerous semi-isolated colonies.

Mayr, on the other hand, tended to view populations, or regionally differentiated subunits of a species, as part of the continuum of phenotypic differentiation. Interactions and differential fates of such groups play no large role in Mayr's theory (though he does speak of Wright's model in several passages). Frequently stressing the moot situation where allopatric and anatomically differentiated "groups" can be variously interpreted as "races" within a single polytypic species (or as newly formed and as yet still allopatric species), Mayr tended to view the significance of quasi-discrete populations as potential incipient species—important in the early stages of the development of "reproductive isolation." It was only later (1954) that Mayr began stressing the importance of small, peripherally isolated populations in the relatively rapid modification of phenotypic and genetic characteristics, leading potentially to relatively rapid speciation events (and thus helping to explain the many gaps reported by paleontologists in the fossil record).

In a context other than Wright's "shifting balance theory," Dobzhansky (1937a, 1941) also saw "races" as incipient species, and his emphasis, like Mayr's, was on the continuum of morphological diversification *and* on the problem of fragmentation of such continuous arrays into discrete species. But "races," even as quasi-discrete subgroups of species, tend in the schemes of both these biologists to be somewhat larger and perhaps a bit more nebulous entities than the semi-isolated reproductive groups (demes) in both Wright's theory and their own renditions and uses of that theory.

In contrast to both Dobzhansky and Mayr, Simpson used the word population but usually in the formal "population genetics" sense (e.g., "population size," one of Simpson's evolutionary determinants). *Subspecies* and *races* appear in the later pages of the book, in conjunction with his discussion of the evolutionary "mode" of speciation. The terms are used almost wholly in the sense of degrees of anatomical differentiation (diversification) along a continuum. Though species may be composed of numerous semi-isolated populations, such a structure has no special significance in Simpson's theory, except, of course, in his early (i.e., 1944) version of quantum evolution, where a relatively small population may undergo rapid loss of adaptation, pass through a preadaptive phase, and finally gain a new set of adaptations within a new adaptive zone (or on a new adaptive peak). It is the population size, rather than the potential differential fates of various demes, portion of Wright's theory that attracts Simpson's attention. Presumably, though he never makes it clear, the relatively small populations undergoing quantum evolution are *portions* of some ancestral species.

Neither Dobzhansky, Mayr, nor Simpson discuss populations in the ecological sense. For the remainder of this book, I will use "demes" when the functional subdivision of species (if they exist) are meant in the reproductive sense. I will use "populations" for such clusters of conspecific organisms that are integrated with other such clusters to form ecological communities or local ecosystems.

It is when we reach species that some interesting and significant differences arise among these three authors. There remains to this date no clear consensus on the "reality" of species, and contemporary views on the issue very much reflect the diversity of views on the overall nature of the evolutionary process. In a nutshell, if one sees species as merely arbitrarily defined collections of organisms united by shared possession of one or more features (i.e., if one sees species as classes), the way is paved for seeing evolution as nothing more—or less—than the transformation of phenotypic (and, of course, underlying genetic) properties of organisms within populations from generation to generation. Or perhaps it is the other way around: such a position on the nature of the evolutionary process demands such a view on the nature of species.

It is, of course, an easy matter to scan these four books of Dobzhansky, Mayr, and Simpson and find ample quotes supporting the contention that all three biologists looked at species in this way—as classlike entities, mere collections of similar organisms. But I have already discussed in detail here (Chapter 3) and elsewhere (Eldredge 1982a) Mayr's equivocation on the subject. Both Dobzhansky and Mayr, it must be remembered, saw the twin themes of evolution as *diversity* and *discontinuity*. By diversity, both meant (for the most part, if not exclusively) *phenotypic* diversity (as opposed to *taxonomic* diversity; not everyone would agree that there is a difference—see Eldredge 1979). Dobzhansky saw the origin of all such diversity in mutation and the basis of all such diversity as genetic. Both he and Mayr saw a continuum, a spectrum in such diversity ranging from slight differences between closely related members of a population, on up through well-differentiated polytypic species, closely related congeneric species, and finally linking up all members of the world's biota.

The basis for such a spectrum of continuous phenotypic diversity is selection—natural selection in the sense of "differential reproductive success" among members of a sexually reproducing species. Selection modifies adaptation; adaptive change is gradual, though it may vary in rate. ("Gradual" seems to have meant smoothly gradational in a descriptive sense as often as it had overtones implying slow or moderate rates.) Both Mayr and Dobzhansky are at pains to insist that not all phenotypic change in evolution is adaptive in nature and thus to be attributed to selection. Several times Dobzhansky (1937a, 1941) says that to believe all such change to be adaptive is "to believe in miracles." Yet the bulk of such change in Dobzhansky's and Mayr's (and in Simpson's) view is most certainly adaptive. And it is here that species, in the theoretical framework of all three authors, are most definitely themselves segments of an evolving continuum. Ayala's (1975) contention that there is a correlation between "genetic distance" and degree of diversification along an axis of among-individual variation on up through reproductive isolation between closely related species fits in exactly with the notion that speciation is but a stage in the "passing stream" of phenotypic–genotypic diversification (once again to quote Mayr 1942). And as we shall see here in chapter 5, Dobzhansky's third edition (1951, pp. 9–10) takes the continuity argument to the extreme in his adaptive explanation of the diversity of all life.

But the issue is not so simple. Even within the context of the (mostly) adaptive origin of the intergradationally diversified biota, there is a distinction between spectra of continuously changing (through time) or varying (through space) suites of phenotypic characteristics and the underlying notion of what a species is. Only when the question of species recognition is broached (not at all in Dobzhansky 1937a and 1941 or in

Simpson 1944, but discussed extensively in Mayr 1942) are the two issues confounded. Species are seen as arbitrarily delineated segments of evolving continua, defined and recognized by some convenient combinations of characters gradationally defined—or serendipitously demarcated by "gaps in the record." But it is important to realize that such temporal conceptions of species, however labile, involved more than a notion of phenotype and genotype; an "evolutionary" species, to quote Simpson (1961, p. 153; see also Wiley 1978) is a "lineage (an ancestral–descendant sequence of populations) evolving separately from others and with its own unitary role and tendencies." In fact, Simpson has long maintained (1961; also as quoted in Mayr 1980c, p. 463) that the "biological species concept," geared as it is to a single time plane, is "nonevolutionary or preevolutionary" as it lacks the temporal dimension—a fault his definition was explicitly designed to remedy. My point here is simply that in the context of diversity it is true that Mayr explicitly (and I believe the same is true for Dobzhansky implicitly) saw species as segments of lineages. The arbitrariness of their recognition, of chopping up such a continuum by some arbitrary set of criteria, reveals the classlike nature (see my chapter 5) of this vision of species (Simpson advocated the same methodological procedures in 1961), *but nonetheless both Mayr and Dobzhansky had a firm concept of lineage when they thought about species through time.* Indeed, their conceptions seem hardly any different from Simpson's. In other words, in thinking of diversity the emphasis is on the phenotype and underlying genotype, and on the phenotype's continuity in transformation. But none of these authors ever lost sight of the fact that at any one point in time the phenotypes are carried by organisms that belong to a single reproductive community, a reproductive community that may persist from generation to generation for a truly long time. In that sense, the temporal view of species implicit in Dobzhansky (1937a, 1941) and fairly explicit (though not stressed) in Mayr (1942) seems to endow species with lateral boundaries (it is forever "reproductively isolated from other such groups," unless it hybridizes) but not with definite bottoms and tops (hence my term *classlike*). In terms of a species' beginning and end, it may indeed be an arbitrarily delineated segment of a continually transforming lineage. The difference between phenotypic transformation and the existence of species as taxa—lineages persisting through time—is not sharply seen by any of these authors.

Simpson (1944), on the other hand, hardly discusses species as units existing at any one time. Indeed, the amount of change involved in "raciation" and speciation is so trivial that Simpson says it must, in general, have little to do with the truly large-scale change that paleontologists can document—and that his theory of quantum evolution was specifically geared to address. Such a view belies a nearly total equation of the process of speciation with "enough phenotypic–genetic transformation to allow a competent systematist to recognize a new species." Yet Simpson too (explicitly in 1961 but also clearly in 1944) was aware of the notion of reproductive isolation, which "has been sufficiently stressed." In point of fact, such isolation played no real role at all in Simpson's theory—a major difference between his views and those of Mayr and Dobzhansky, and a difference that reflects something more than the obvious fact that each of the three was starting with a different data base.

Dobzhansky (1937a, 1941) saw discontinuity, the second of the twin themes of evolution, as arising from two sources: at the level of the gene (genes are particulate) and through the reproductive isolation of species. For his part, Mayr spoke of "bridgeless gaps" between species. The discontinuity, the gaps between species, are both phenotypic and, of course, reproductive. The "isolating mechanisms" keeping species separate have a genetic basis and are phenotypic at least in the most general sense. So the reproductive aspect of discontinuity is caused by genetic–phenotypic differences which are then enhanced (the gap widens) once reproductive isolation is complete. The phenotypic differences involved in the gaps originate in the normal processes of diversification and are thus usually adaptive in origin. Mayr and Dobzhansky, at least in their early works, differed sharply on the pros and cons of an adaptive basis for the origin of reproductive discontinuity. Dobzhansky struggled mightily to establish such an adaptive basis, which Mayr did not require at all in his theory. As will become clear below, such a difference is by no means trivial, as it has deep ramifications for the integrated theories of both; though Dobzhansky and Mayr agreed substantially about what species are and how they evolved (to the point of indulging in low-key banter about which one of them actually came up with the "biological species concept"), the two differed on the more general issue of why there are species in the first place and what is their true role in the evolutionary process.

Both Mayr (1942) and Dobzhansky (1937a, 1941) see species formally as reproductive communities. Mayr details the long history of this view, but it is fair to say, I think, that it was not really until both these biologists emphasized the "reproductive community" nature of sexually reproducing species that evolutionary theory in general formally embraced the origin of species as a topic worthy of serious consideration. In this light, species are at least partially—and potentially totally—discrete entities in nature. And their cohesion comes from a network of reproductive interconnectedness, *not* from the joint possession of a common suite of phenotypic characters with an adaptive basis. Simple reduction to the processes governing adaptive change within populations becomes thereby difficult, if not logically impossible. To Mayr (1942), enforced geographic isolation is a prerequisite for the formation of such discrete reproductive communities, a view he has always staunchly maintained. Selection by itself cannot fragment a reproductive community, because selection works to modify adaptation gradually. It alone cannot give rise to discontinuity. Dobzhansky largely agrees about the importance of geography. But Dobzhansky is rather like the extreme selectionist R. A. Fisher, whom he cites. Dobzhansky (1937a, 1941), as we have seen, was determined to show

how selection can create reproductive discontinuities. To him, there is a strong, adaptive, *genetic* reason why we have species at all: species maximize the focus on adaptive peaks and minimize the realization of all those potential inharmonious gene complexes. Since the existence of species is adaptive in the first place, for the biota to be assembled into reasonably discrete reproductive clusters it is only natural to assume, as Dobzhansky did, that their formation is achieved by natural selection.

As Paterson (1978, 1981, 1982; see also chapters 6 and 7 herein) has pointed out, both Mayr and Dobzhansky emphasized reproductive isolation from other species, rather than the "positive" aspect that the mechanisms serve mainly to allow potential mates within existing species simply to recognize each other and not to keep them from wasting gametes trying to hybridize with members of other species. Paterson concludes that it makes no sense to speak about selection *for* "reproductive isolation"; such isolation is a mere byproduct of selection *for* mate recognition. The point is well taken (see my further use of it here in chapters 6 and 7). However, we should remember that Dobzhansky's and Mayr's stress on isolation arose in the context of the explanation of discontinuity, and from this perspective they were merely asking how a continuous array becomes segregated into discrete (i.e., isolated) clusters. And Mayr, at least, used the term *isolation* mostly descriptively and certainly not primarily as the outcome of selection, as we have seen here in chapter 3.

According to Dobzhansky, not only do species exist in some concrete sense in the context of reproduction (and not, once again, in "diversity"), but they also have a role in evolution. They stay on and track adaptive peaks. Progressive division of such peaks—clear in the passages cited (Dobzhansky 1941, quoted here in chapter 2)—leads necessarily to a nested array of adaptations and taxa. But it is the (at least potentially) ever-changing lineage view of species that Dobzhansky has in mind here; thus, his view of monophyletic taxa is very much one of lineage transformation (at variable rates, presumably) and lineage splitting. Monophyletic taxa—just like his temporal view of species—are distinct laterally (they occupy different peaks) but smoothly connected intergradationally through time, in time-honored Darwinian transformational fashion.

Mayr, on the other hand, sees a different role for species in evolution. He characterizes the "new systematics" as concerned with variability within species and not so much with species per se, and there is relatively little in Mayr's 1942 book about the actual role species might play in evolution. In contrast with his later position (e.g., Mayr 1963, p. 621; see Eldredge and Cracraft 1980, p. 275), Mayr sees no special relationship between species as such and macroevolution in 1942. The concluding lines of his book see macroevolution rather as an "extrapolation of speciation," and he states that "all the processes and phenomena of macroevolution and the origin of the higher categories can be traced back to intraspecific variation" (Mayr 1942, p. 298). Here the continuum of (morphological) diversity is stressed completely over the possible significance of discontinuity. In 1963, though still somewhat equivocal, Mayr stressed the discreteness of species and the consequent implications of such discreteness for macroevolutionary theory. Developing the idea that speciation may supply the raw material for macroevolution (just as mutations do for natural selection at a lower level; cf. Wright 1967), Mayr writes (1963, p. 621): "Species, in the sense of evolution, are quite comparable to mutations." Though nothing really like this view is present in his earlier work, nonetheless it is foreshadowed—as indeed is the entire concept of species as individuals (Ghiselin 1974a; Hull 1976)—in part of his discussion bearing on the reality of species (see chapter 3 herein).

In sum, in the three books where diversity and discontinuity are identified as the twin central themes and problems of evolution requiring simultaneous explanation—Dobzhansky 1937a and 1941 and Mayr 1942—the answer to What is a species?" depends upon the context, that is, which of the two aspects of evolution is uppermost in the author's mind. In terms of the perceived continuum of diversity (which, again, means genetic–phenotypic diversity), species have "lateral borders" (there are morphological gaps between species, however slight they may be in some cases), but through time they have no discreteness, no particular beginnings or endings (unless extinction terminates the lineage). From this limited perspective, species have the attributes of classes, not individuals (see my chapter 5 for more on this). Moreover, this seems to be Simpson's entire view of the nature of species. In the context of discontinuity, however, species are definite entities, even though, as Mayr says, allopatry may raise troublesome epistemological problems of definition and recognition. The early synthesis is equivocal on the ontological status of species. And the same spectrum of opinion on the nature of species is still very much with us today.

I have discussed the confusion, general in comparative biology, between taxa and categories in several footnotes to chapters 2 and 3. The distinction between actual historical entities (Charles Robert Darwin, *Tyrannosaurus rex*, Animalia) and categories (organisms, species, kingdoms) is crucial (see chapter 6 herein). *But the fact that all three of the architects of the synthesis routinely called taxa "categories" is not the basis of the claim that they treated such entities as classes rather than as individuals.* As I have already remarked, it is appropriate to read "taxon" wherever "category" is written in these works.

But we still need to know what sort of ontological status these writers saw for taxa of rank higher than species. And this is extremely difficult to specify with anything resembling precision. Simpson's entire book, I have argued, is basically geared to the development of a theory of megaevolution—his "quantum evolution"—which "is, however, believed to be the dominant and most essential process in the origin of taxonomic units of relatively high rank, such as families, orders, and classes" (Simpson 1944, p. 206). In the preceding sentence, Simpson says that "like the other modes, it can give rise to groups of any size, and the sequences involved can be (subjectively) divided into *morphological units of any desired scope, from subspecies up"* (italics added). I can only conclude that Simpson's view of species holds for taxa of all rank on up the Linnaean hierarchy; larger chunks of lineages, with relatively more change accrued, are classified relatively higher than subspecies and species, which are the smallest chunks of these lineages, these evolving continua. Taxa delineated this way, according to some degree of differentiation or transformation, are classlike and certainly not entirely discrete pieces of biotic "furniture." As we have seen, such too is the prevailing view (insofar as can be determined) of Dobzhansky (1937a, 1941) and Mayr (1942). Perhaps the most intriguing point of all concerning these early treatments of taxa of rank higher than species is that they are hardly treated at all. They are shadowy things indeed. To all three of these authors, macroevolution seems far more to encompass degree and kind of phenotypic change— change that allows us to define and recognize taxa of relatively high rank, to be sure. But "the origin of taxa of higher categorical rank" seems to have been virtually synonymous with "origin of major new adaptive types" in these earlier works.

And finally, if only for the record, none of these three authors supplies an explicit discussion of what communities, ecosystems, or regional biotas might be, or what roles they may play in the evolutionary process— despite Dobzhansky's indignant protestation that it is nonsense to say that genetics pays no heed to the environment. The ecological side of things plays no *explicit, formal role* in either structure or content in the evolutionary theories of Dobzhansky, Mayr, and Simpson, except in the conceptual nexus that sees natural selection forging adaptations (phenotypic attributes of organisms) to maintain, improve or maximize fitness. And, of course, it is the environment—largely abiotic but also seemingly biotic— that is the source of such selection forces.


## EVOLUTIONARY PROCESSES AND PATTERNS
## IN THE EARLY SYNTHESIS

*Process*, of course, is a slippery word. Some evolutionary biologists of late (e.g., Cracraft 1982) have suggested that even such a well-known evolutionary process as natural selection is really a *pattern*—a systematic phenomenon of differential reproduction. Allopatric speciation likewise is a class of events, a common pattern of fragmentation of reproductive communities, of which several distinctive subpatterns (as in Bush's 1975 modes a and b) can easily be delineated. Nonetheless, natural selection, allopatric speciation, and other examples seem to retain a legitimate flavor of process. Natural selection is the mechanism, deterministic albeit statistical, that has been abundantly shown to underlie many cases of generation-by-generation change of allelic frequencies within populations. It is the nexus between the genetics of populations and the external milieu. Similarly, allopatric speciation, with its twin aspects of diversity and discontinuity, invokes a range of ecological and organismic processes that transform one coherent sexually reproducing species into two.

But admitting that allopatric speciation embraces a host of such ecological and organismic processes opens a Pandora's box in a section on processes in a critical analysis of the early synthesis—or, for that matter, in any evolutionary theory. This is because evolutionary theory *is* a theory of process. That is its natural content. To list here, and to compare and contrast, all the processes that Dobzhansky, Mayr, and Simpson saw as relevant to the evolutionary process in toto would be futile and largely redundant on the discussions of the preceding two chapters. Nonetheless, a brief discussion is necessary if only to classify what major sets of processes each author thought were most relevant to a complete evolutionary theory. It is their choice of processes, the relative importance assigned to each, and the levels on which those processes worked that not only outline the contents of the theory of each of these three zoologists but give the structural outlines of those theories as well.

There is another reason to consider pattern and process together in this section: the classes of historical events that form the basis of our recognition of the very patterns of evolution—the adaptive radiations, long-term directional trends, clines, balanced polymorphisms, and so on— are the very phenomena to be explained by a theory of process. Dobzhansky (1937a, 1941) was perhaps the most emphatic and explicit about the connection between theories of process and their empirical support in the laboratory and in the field. We need pattern to verify (current usage prefers "test") notions of process. Thus, the intimate relationship between pattern and process in evolution is a two-way street. Grene (1959) has argued that the connection is but one-way, at least in Simpson's work, in the synthesis—that the data are simply explained away by theory, which itself remains relatively immune to criticism. (See Teggart 1925 for a penetrating analysis of Darwin's treatment of the fossil record, which comes to much the same conclusion.) Grene's criticism of Simpson provides an entry point to an understanding of where the synthesis falls short of formulating a complete, and completely useful, evolutionary theory.

Simpson's (1944) evolutionary theory is in an important sense hierarchical. There are classes of events, three in number, which are the modes of his *Tempo and Mode*: speciation, phyletic evolution, and quantum evolution. This tripartite characterization of evolutionary phenomena has hierarchical overtones because of the connection Simpson draws between them and the ranks of the Linnaean hierarchy. As we have seen, speciation to Simpson entails adaptive differentiation within species ("raciation"). When and if enough such adaptive change accumulates within a lineage, when the morphological change amounts approximately to the degree of difference we typically see between two closely related modern species, we say that a new species has evolved. Simpson (1944) shows that he is well aware of allopatric speciation, though this process plays no formal role in his evolutionary theory.

Thus, Simpson's hierarchy of process is not nested. It is a ranking, like the chain of command in the armed services. As with his earlier (i.e., in the same 1944 book) trichotomy of microevolution, macroevolution, and megaevoloution, there is an implication that somehow the higher-level processes are more important than the lower—as we saw in Simpson's remark that what happens to a billion rats over ten million years is "obviously more important" than what happens to a hundred rats in ten years.

Taxa of various ranks are complexly internested. It is only the ranking of the categories that forms a linear array. The hierarchical array of taxa is isomorphic with the complexly nested hierarchies of homology (chapter 7 herein). It is important that, although Simpson's phyletic and quantum evolution modes are addressed to the taxa above the rank of species, the hierarchical aspects of his theory are not addressed to either the taxic or the homology hierarchies. As we have seen, Simpson saw taxa—from subspecies up through orders, classes, and phyla—as greater or lesser segments of evolving lineages. In short, evolution to Simpson is almost purely a matter of the transformation of phenotypic (and underlying genotypic) properties of organisms.

And that is why quantum evolution, phyletic evolution, and speciation are "modes," not "processes," in Simpson's presentation. Simpson is quite consistent throughout his book; the determinants of evolution lie entirely within the province of genetics for their experimental study. They include such processes as mutation and natural selection, but also such parameters as mutation rate, population size, and generation time. It is the combination of these determinants—in various degrees and intensities—that yields his three modes. But the list of evolutionary determinants is brief indeed. As I concluded in the preceding chapter, Simpson in 1944 was no straightforward reductionist, but that is so only because he sees the same fundamental ingredients put together in three basically different ways to produce his three different modes. In terms of basic classes of evolutionary processes, Simpson *is* more of a reductionist than either Mayr or Dobzhansky when we compare the contents of these early works. Later blurring of the distinction between quantum and phyletic evolution yields the paradigmatic reductionist central thesis of the synthesis as we know it today: macroevolution—however defined but involving the "origin" of "taxa of higher rank" (and, implicitly, relatively larger-scale phenotypic modification or innovation)—is simply microevolution writ large. Simpson did mention (at least twice) multilevel selection and discussed Stufenreihe briefly and in such a manner as to suggest phenomena of differential group "success," but such discussions are only minor disruptions of what is otherwise a markedly consistent and persistent theme of adaptive transformation.

Dobzhansky's theory, on the other hand, was not so simple. As I pointed out in chapter 2, Dobzhansky explicitly characterized his theory in hierarchical terms. And, like Simpson, Dobzhansky was utterly consistent throughout his text; the structure of his book reflects his hierarchical view of the evolutionary process.

The evolutionary process is hierarchical because those elements of the biotic world involved in the evolutionary process—genes, organisms, demes, species, and monophyletic taxa—define a series of levels. Unlike the present development of the genealogical hierarchy (chapter 6 herein), Dobzhansky saw no particular significance in the inclusive, nested nature of this particular hierarchy, though he was clearly aware that genes are in organisms, organisms form demes, and demes form species. As we have seen, discontinuity arises at the genetic level and again at the species level in Dobzhansky's theory.

But the levels that Dobzhansky sees are also extremely important to his analysis of the evolutionary process. Perhaps his most fundamental, explicitly hierarchical view here remains standard in evolutionary thought: genetic mutations occur in organisms and are of great importance as the ultimate source of diversity in evolution. But mutation is a (biochemical) process within organisms and the province of study of "physiological genetics." If left somehow at that level, mutations would have no evolutionary impact, because organisms themselves do not evolve. It is only when mutations are injected into the gene pools of populations that they take on evolutionary significance. And Dobzhansky is eloquent when he tells us that the rules of the genetics of populations are different from those of "physiological genetics."

This two-level vision—genetics of individuals and genetics of populations—is a core aspect of evolutionary theory to the present day. It is, or should be, understood by all beginning students. It has been examined in detail by philosophers of science, geneticists, and population biologists alike. However, the usual context in which this point is retained and utilized in present-day discussions is somewhat narrower than Dobzhansky's hierarchical scheme. Specifically, it is common to read that mutations are random with respect to the regime of natural selection to which they are exposed. This is an axiom of contemporary theory. Variation arises at a lower (genic within organism) level, and selection operates on the variation presented to it within the higher (among-organism, within-population) level. Dobzhansky, of course, saw the relation between the processes of mutation and selection this way, but his explicit invocation of levels was more general. All processes leading to heritable phenotypic innovations are intracellular, hence "level one." All such innovation is injected into the population level. And all "sortings" of such variation—selection, but also the sampling errors of drift—are second-level phenomena. Explicit recognition of such levels leaves the door open for discussion of additional processes contributing to such variation and sorting (see Vrba and Eldredge 1984 and chapter 6 of this book for a discussion of lower-level introduction of variation and of sorting at various levels of the genealogical hierarchy).

Dobzhansky does not go beyond Wright in his elaboration of intra- versus interdemic processes. Demes figure largely in their positive and negative effects on net standing genetic variation within a species. Interdemic selection, in fact, seems to take a back seat to the stochastic aspects of interdemic survival in Dobzhansky's (from Wright 1932) view of the evolutionary process. But demes do constitute a distinct level in Dobzhansky's theory.

It is at the level of species that Dobzhansky once again sees discontinuity arising. Not as adamant as Mayr (1942) that speciation must be allopatric, Dobzhansky nonetheless concedes that reproductive isolation between two (or more) divisions of a species usually is geographically based to start. Dobzhansky and Mayr differ here, mainly because Mayr sees no possibility for selection sundering the coherence of a reproductive community, whereas, as we have seen (chapter 2), Dobzhansky strove mightily in his first and second editions to show that reproductive isolation is adaptive—that there is a reason why species are reproductively isolated (his reason: to prevent the formation of inharmonious gene combinations suitable only for adaptive valleys), and thus under the aegis of natural selection.

Dobzhansky's hierarchy of biological levels involved in the evolutionary process stops with species. He sees no special role for species within monophyletic taxa in macroevolution. Indeed, he (1937a, p. 12) puts a "sign of equality" between the processes of microevolution and macroevolution. But he still thinks hierarchically in conjunction with macroevolution, as in his use of the word *hierarchy* in alluding to what I call the hierarchies of homology (chapter 7 herein). The complex hierarchically arranged arrays of evolutionary novelties (adaptations, synapomorphies) seem to Dobzhansky to fall directly out of a simple extension of the metaphor of the adaptive landscape. But the interrelated hierarchies of homology and taxa (the Linnaean hierarchy) are products of the evolutionary process and not at all the same thing as the genealogical hierarchy—the term used by Eldredge and Salthe (1984) and Vrba and Eldredge (1984) for their expanded version of the gene–organism–deme–species hierarchy that provides the structure of Dobzhansky's theory.

To put it baldly, then, the main sets of processes that Dobzhansky discusses in his books are mutation (in a very broad sense), recombination, selection, drift (to a lesser extent than selection), and the origin of isolating mechanisms—that is, speciation. But it is his hierarchical view of biotic levels and where and how those processes enter in that gives his theory its unique flavor. Evolutionary theory, in general and until recently, has lost sight of much of the explicitly hierarchical view Dobzhansky developed in his *Genetics and the Origin of Species*. It has become the poorer because of this loss.

Mayr (1942) focused very much on the same general set of processes as did Dobzhansky. While Dobzhansky devoted considerable space to formal demonstration that mutation ultimately underlies all diversity, relied heavily on Wright for a view of the genetics of populations, and then contributed heavily himself to a discussion of the origin of reproductive discontinuity in nature, Mayr focused on the issue of the adaptive nature of patterns of phenotypic diversification in nature, before turning in his book's second half to consider patterns and processes in the origin of reproductively isolated taxa—new species. That his book is no mere repetition of Dobzhansky's should be obvious from my preceding two chapters, yet in summarizing the two authors' evolutionary theories as developed in their early books, one finds more of a difference in emphasis than in choice of which processes are active in evolution. Thus, Mayr thinks that mutation is the ultimate source of variation, that selection effects genetic change in populations, and that much if not all phenotypic variation both within and among species is adaptive in nature. As just noted, Mayr sees reproductive discontinuity between species essentially as an accident of geography—the accumulation of genetic differences between geographically isolated groups. Such genetic differences represent normal genetic modification within two (or more, of course) groups no longer able to exchange genes. Should such genetic differences accumulate to the point where reproduction among the members of each group is fully impeded (should sympatry be reestablished), speciation has occurred. Partial impedance raises a number of possibilities, and Dobzhansky's notion of reinforcement is at least partially appealing to Mayr in 1942.

Mayr does not take an explicitly hierarchical stance in *Systematics and the Origin of Species*. Indeed, as I have noted, part of his argument (especially in his first hundred pages) emphasizes continuity (in morphological diversity), from genes through species and beyond, and the existence of levels is if anything submerged in favor of a reductive, adaptive theme so familiar in the the modern synthesis. Yet, as I also noted, it is Mayr's treatment of the ontological status of species—where, at least in some contexts, species are seen as "real," concrete entities—that has led to his own, and others', view of a complex role that species themselves may play in macroevolution. In 1942, Mayr's view of macroevolution entails no process of species selection—or any set of evolutionary processes other than those of microevolution. But in an important sense his views in 1942, in their later development (by himself, but also by others, as detailed in chapter 5 herein), lead the way to an expansion of Dobzhansky's hierarchy.

If it is difficult to analyze and characterize Dobzhansky's, Mayr's, and Simpson's books accurately, as I have tried to do in the preceding chapters, it is an even more difficult task to summarize and compare their theories briefly without subjecting them to caricature. The hierarchical view of evolution and evolutionary theory developed in the second half of the present book takes as its springboard a definite view of the ontological nature of various biological entities and discusses pattern and process—their existence as well as their study—explicitly within such a hierarchical framework. That is why, in my summary, I have focused on the ontology and structure of the theories of these three founding fathers. But the content of their theories and the complexities, doubts, and uncertainties that each strives to acknowledge fall by the wayside in such a brief encapsulation.

I am also aware that the present "summary of a summary" is rapidly converging on wholehearted agreement with Mayr's one-two punch characterization of the synthesis (quoted in my chapter 1). Mayr's statement leaves out a lot—literally volumes—but for a few lines it is about the best distillation of these early works imaginable.

If one could superimpose these four volumes on one another, the major differences among them would readily show up as a mere function of interest and expertise among the spectrum of entities and processes of evolutionary import. I have strived to establish the commonality of the works but also the genuine differences that do exist among the viewpoints of these three biologists. It is a temptation, despite my earlier intentions, to analyze the sequence of intellectual events that led to what Gould (e.g., 1980b) and others have seen as a narrowing and hardening of the synthesis—largely, I think, a matter of smoothing out what differences are apparent among the three early architects. But I shall not yield to that temptation. The job has been done already (see the volume edited by Mayr and Provine 1980), at least well enough to establish to my satisfaction that, if anything, Mayr's single-sentence precis of the synthesis is even more accurate when applied to later developments. I will confine myself to the relatively few innovations—mostly in thinking but also from the empirical realm—that have affected the structure and content of mainstream evolutionary thought.

## THEMES AND VARIATIONS WITHIN THE SYNTHESIS

What lies within, and what beyond, the boundaries of the synthesis depends very much on the eye of the beholder. A narrow conception would see some of the themes—such as "non-Darwinian evolution" (neutrality) and the possibility that selection goes on at more than one level—as lying outside the purview of the synthesis. Yet both concepts are found—if not prominently featured—in some of the four books I discuss in chapters 2 and 3. On the other hand, an excessively loose version of the synthesis effectively says that all evolutionary theory (presumably, and so long that it contains the core neo-Darwinian paradigm of genetically based interorganismic variation and selection) falls within the scope of the synthesis—a reasonable characterization, I think, of the Stebbins and Ayala (1981) position.

This caveat sets up my apologia for including here (1) grades and polyphyly, (2) neutrality, and (3) levels of selection as three major themes and variations that have developed since the synthesis was born and did its "maturing" or "hardening." I feel that, whereas the roots of all three of these themes lie very much within the founding documents (and, indeed, lie far deeper in the recesses of the history of biological thought), these three topics have occasioned renewed vigor and intellectual ferment. To some extent, and in various ways, these topics have highlighted slight shifts in both structure and content of the way evolutionary biologists think about evolution—the central theme of this book.

Yet I include my discussion of these and one or two additional topics within a chapter that strives to capture the essence of the synthesis. My criteria for doing so are simple enough: within the synthesis, there was early on—and continues to be—a dominant view on the range of entities deemed to be of evolutionary importance, together with a set of views on the ontological status of such entities. I have reserved for later discussion any argument or data set that either expands the range of those entities (genes through monophyletic groups) or causes a radical shift in our onto-logical view of such entities, as in Ghiselin's (1974a) view of species as individuals. I find such topics ineluctably leading—sometimes quite explicitly—to a hierarchically structured evolutionary theory, similar to but more extensive than the one invoked by Dobzhansky in 1937. In con-trast, the themes and variations I discuss here fall well within the range always included within the synthesis and occasioning no radical shift in our concept of the nature of the entities—genes, organisms, demes, spe-cies, and monophyletic taxa—away from the views already collectively expressed by Dobzhansky, Mayr, and Simpson.

## Grades and Polyphyly

J. S. Huxley's (1958) paper on grades and clades was the touchstone for a number of research articles and reviews on macroevolution. Its influence reached well into the mid-1960s. Evidently expressing an idea whose time had come, Huxley pointed out that there are two ways of defining and recognizing collections of species—monophyletic groups (which he called clades) and sets of organisms united by the possession of the same (or apparently the same, or at least highly similar[1]) anatomical, physiolog-ical, and behavioral adaptive properties that constitute a stage of evolu-tionary advance:

I accordingly suggest that the customary terminology purporting to define phy-letic units should be supplemented by a secondary terminology aimed at delimit-ing steps of anagenetic advance. As will be made clear later, such units of anage-netic change are just as "natural" or at least non-arbitrary as the customary monophyletic units, such as genus, family, class, etc., and can be readily recog-nised and delimited. The best general term for such anagenetic units would seem to be *grade*. . . . I further suggest the term *clade*, to distinguish monophyletic units of whatever magnitude. (Huxley 1958, p. 27)

According to Huxley, recognition of both such kinds of groups—taxa defined and recognized both ways—takes cognizance of the dominant processes of evolution:

This leaves us with improvement, diversification, and persistence as the three main types of evolutionary process. Since precise scientific definition requires the use of technical terms, I shall use *anagenesis, cladogenesis,* and *stasigenesis* to denote these three types of process. (Huxley 1958, p. 23)

A number of evolutionary biologists, predominantly paleontologists, adopted Huxley's gradal terminology. Many taxa were found to be grades rather than clades: the goniatites, ceratites, and ammonites among Ammonoidea; the chondrosteans, holosteans, and teleosteans within Osteichthyes (Schaeffer 1965); the various groups of placoderms (Miles 1969); and so forth. "Experimentation" (Schaeffer 1965) and "multiple evolutionary pathways" (Bock 1959) were among the expressions coined to denote modes and means of adaptive change that lead to new grades, or "levels of organization" (Hecht and Schaeffer 1965). It was widely claimed (e.g., Simpson 1959b; Schaeffer 1965) that the attainment of new gradal levels was often achieved by a number of separate lineages emerging independently from the ancestral grade; thus, the theme emphasized the polyphyletic origin of such groups. Indeed, Coon (e.g., 1962) saw the species *Homo sapiens* as a grade that arose separately a number of times, corresponding to the major racial divisions of our species.

What are these grades? Huxley (1958) saw them as occupying all levels of the Linnaean hierarchy, citing examples of horse genera (based on a letter from Simpson) on the one hand and truly large-scale groups, such as Metazoa, on the other hand. Paleontological data are, in my own experience, fraught with examples of species, genera, families, and so forth, that succeed one another stratigraphically (i.e., through time) and seem to record some form of progression, if not always a strictly linear trend. Such series are, in my opinion, not yet fully understood or explained (but see Ch. 7 for further discussion). For a cladist, it is easy to assert that taxa alleged to be monophyletic in such series cannot in fact be monophyletic—hence, the taxa are in an important sense not "real" (cf. Eldredge and Cracraft 1980). Yet the fact remains that *anatomical* progression is frequently arranged in such time-successive clusters. Thus, grades do seem to exist in the sense that Huxley describes.

It is important to stress that grades have some apparent, if as yet vague, empirical basis if only because it is tempting to point to the popularity of grades and polyphyly as the triumph of theory over data. For there is no doubt that grades and polyphyly (as explicit positions on the ontological status of "higher categories" and the processes underlying their origin) reflect a deep commitment to adaptation as *the* fundamental phenomenon in evolution. Gone perhaps is the imagery of adaptive peaks and zones, but instead there is a sort of gross-level "optimality" argument—one parallel to optimization arguments still very common in the literature of population biology and within-species evolution. Limited for the most part only by the constraints of history (i.e., the adaptive configuration of the ancestral grade with its underlying genetic machinery) and the constraint of architectural limitation of design, evolutionary progress results from improvements in adaptation—perhaps by invention of "key innovations" (Simpson 1959a) of varying degrees of complexity, or the elaboration of new bauplans, both of which have been seen as potential triggers for "adaptive radiations."

Thus, anagenesis, grades, and polyphyly represent simply an intensification of the theme of progressive adaptation as perhaps the central phenomenon of evolution and certainly the dominant (if not the sole) ingredient of large-scale (macro-) evolutionary change.

Huxley (1958, p. 27) said that grades "are just as 'natural' or at least non-arbitrary as the customary monophyletic units"—and herein lies the greatest significance of grades to the present inquiry. There is a world of difference between "natural" and "non-arbitrary"—at least insofar as I construe the terms in an evolutionary context. "Natural" would seem to imply "real," the existence of a unit—an entity—in nature. Such entities (the individuals discussed more extensively in ensuing chapters) have, minimally, (1) beginnings, histories, and endings (they are spatiotemporally bounded), and (2) some sort of "glue" that imparts cohesion and contributes to their boundedness. "Non-arbitrary" means simply that we can recognize something. To belong to a grade, an organism needs only to possess the requisite characters. Such collections need not be spatiotemporally bounded, though they may be. Reptiles are a nonmonophyletic grade that range from the Carboniferous to the Recent. But, more importantly, the sort of genealogical connectedness (ongoing speciation) that provides the glue that indeed keeps a monophyletic taxon of rank higher than species "alive" is simply missing in a nonmonophyletic grade. Grades are—and were really only intended to be—morphological abstractions. A taxon is defined and recognized as a grade, but the emphasis and the theoretical interest attached to such recognition is the set of morphological features that define the adaptive complex that *is* the grade.

Thus, grades have the classlike nature of "higher categories"—very much the views of Mayr, Dobzhansky, and Simpson in their early books, but now made far more explicit in Huxley's terminology. And if grades are nothing but constellations of morphological features of organisms that confer some adaptive specificity or other, it is clear that no theory other than within-population natural selection need be adduced as explanation of their existence. As Simpson still felt in 1959a (p. 255), "the basic processes are the same at all levels of evolution, from local populations to phyla, although the circumstances leading to higher levels are special and the cumulative results of the basic processes are characteristically different at different levels."

## Multilevel Selection: The Synthesis Copes

Many of the recent discussions of hierarchy in evolution seem almost to synonymize hierarchy with selection at different levels. A fundamental premise of this book is that the biological entities that participate in the evolutionary process are themselves hierarchically arranged in a nested fashion. Thus, ontology virtually forces our evolutionary theory to be structured hierarchically. In fact, there seem to be several such hierarchies (Eldredge and Salthe 1984; chapters 6 and 7 herein). Selection may or may not go on in some or all of the levels of one or more of these hierarchies.[2] Indeed, Dobzhansky's (1937a) earliest fully articulated evolutionary theory was explicitly hierarchical, yet he saw selection only at the among-organism, within-species (or within-colony) level.

Yet multilevel selection—that is, selection that is something other than differential reproductive success among organisms within a deme—obviously does imply hierarchy, even if the relations among the various sorts of entities supposedly being selected are not specified in detail. And it is obvious that if an entity is to be selected among others of like kind, it must in some material way be said to exist. It cannot be a class of things, a set with members, but rather it must be a whole with parts. It must be a historical entity.

Thus, discussions that explicitly invoke selection at more than the single conventional level definitely herald the more full-blown hierarchical approach to evolutionary theory nowadays emerging. Some of these are more appropriately reviewed in ensuing chapters where I take up hierarchy in detail. But it is an explicit and intransigent resistance to such ideas that stamps the bulk of evolutionary thought since the core books of Dobzhansky, Mayr, and Simpson appeared and that falls squarely within the purview of the synthesis. Advocacy of selection at various levels is generally taken as misguided departure from orthodoxy. And far more interesting than mere defensiveness, reactions against heterodox notions of higher-level selection led directly to the hyper-reductionism of such notions as Dawkins's (1976, 1982) "selfish gene." As I shall develop, in a way such extremes in argument merely follow some of the themes of the synthesis through to their fullest logical extent. But in focusing attention on evolutionary processes at the ontological level of the gene, the last vestiges of Dobzhansky's embryonic hierarchical scheme were utterly removed—and with them went unexamined some of the problems that troubled Dobzhansky.

Figuring prominently in the ongoing "levels of selection" debate is the work of G. C. Williams, particularly his 1966 book *Adaptation and Natural Selection*. Subtitled "A Critique of Some Current Evolutionary Thought," the book seems most often remembered as a staunch defense against the model of group selection propounded by V. C. Wynne-Edwards (e.g., 1962) a few years earlier.

But Williams (1966) did not rigidly dismiss all forms of group selection out of hand. Accepting Lewontin's (1962; Lewontin and Dunn 1960; see Williams 1966, p. 117) example of the *t* locus in house mice as a probably valid example of group selection, Williams merely argued that there would be no need for an onerous, higher-level process such as group selection if it were possible to show that simple natural selection alone could suffice to produce a given phenomenon. Indeed, Williams thought that higher-level selection was a virtual truism in evolution: "Like the theory of genic selection, the theory of group selection is logically a tautology, and there can be no sane doubt about the reality of the process" (Williams 1966, p. 109). Rather, the question in Williams's book is the relative importance of such higher-level (demic, species) selection processes vis-à-vis pure natural selection. And it is accurate, I think, to characterize William's position on higher-level selection throughout his 1966 book that it is a vastly less significant process than natural selection. Williams repeatedly (e.g., 1966, pp. 101, 111, 124) demands that empirical evidence for the existence of higher-level selection include the demonstration that the higher level must operate *against* natural selection, that traits are selected because they benefit the group even though they may be disadvantageous to organisms:

We must always bear in mind that group selection and biotic adaptation are more onerous principles than genic selection and organic adaptation. They should only be invoked when the simpler explanation is clearly inadequate. Our search must be specifically directed at finding adaptations that promote group survival but are clearly neutral or detrimental to individual reproductive survival in within-group competition. (Williams 1966, p. 124)[3]

This itself is quite an onus to place on group selection (as Wimsatt 1980 has pointed out), though it is true that the pragmatic demonstration of group selection is easiest under such circumstances (Alexander and Borgia 1978). Eldredge and Cracraft (1980, pp. 285 ff.) provide several apparent examples where intragroup change seems to some extent to be opposed to longer-term net intergroup change. And I shall follow Vrba's argument (e.g., Vrba 1984b; Vrba and Eldredge 1984) that true group selection in any case must act on true group attributes (actually, "higher-level individual" is preferable to "group"; cf. Hull 1980). And these attributes, of course, cannot be the same as the phenotypic attributes that constitute the adaptive features of organisms.

But from the viewpoint of such phenotypic organismic adaptation (which serves as the prime focus of his entire book), Williams is clear that, whereas "such matters of extinction and survival are extremely important in biotic evolution" (p. 121), the role of biotic adaptation through group selection is on the whole "as a creative evolutionary force that supplements genic selection." (This phrase occurs on p. 124, but not with the conclusion I am attributing to Williams.) The remainder of his book is a skeptical review of all such putative biotic adaptations, including "adaptations of genetic systems," "reproductive physiology and behavior," and "other supposedly group-related adaptations."

*Creative*, of course, is a loaded word. Whereas Williams does discuss many supposed examples of biotic adaptation (e.g., to pick but one at random, "death from old age"—Williams 1966, p. 225), it is clear that creativity in evolution to Williams means the emergence of genic–phenotypic novelty, that is, the sorts of attributes of organisms that constitute his "organic adaptations." For instance, Williams writes (p. 110): "However, it is the origin of complex adaptations, for which the concept of functional design would be applicable, that is the important consideration."

Here, of course, Williams is in good company. For example, Simpson discusses selection at "higher and lower levels" at least twice in his 1944 *Tempo and Mode*, concluding in one passage that higher and lower levels of selection are merely eliminative, and only natural selection, acting among individuals within populations, is "creative" ("an originating force"—Simpson, 1944, p. 31). Natural selection forges new combinations of genes that underly the production of "new" adaptations or the modification of old ones.

Williams simply defends the view that Simpson articulated, and, indeed, from the standpoint of the physical and behavioral attributes that we observe in organisms, the point seems to me inarguably correct. But problems arise, stemming, as Dobzhansky remarked, from the propensity of some biologists to synonymize *adaptation* with *evolution* (e.g., Dobzhansky 1937a, on the views of R. A. Fisher). Though Williams avoids such an extreme position, there is no doubt that he sees organismic adaptation as the central problem or theme in evolution throughout his 1966 book. Thus, discussing ecological interactions in terms of the "effective strategies" of the fox and the rabbit involved, Williams is led to such statements as (p. 68): "The goal of the fox is to contribute as heavily as possible to the next generation of a fox population." It is the further implication of such remarks rather than the specific point Williams was making that interests me here. It is no very great leap from this view of the purpose of organic adaptations to the views of Dawkins (e.g., 1976, 1982), where hens are literally an egg's way of making another egg—the eggs being the nutritive package surrounding the zygote, the cells being the protective housing for the "immortal coils." Indeed, in *The Selfish Gene*, Dawkins (1976) explicitly acknowledges the roots of his views in Williams's work. So, in his well-known exposition of the "selfish gene," evolution is to Dawkins (p. 48) "the process by which some genes become more numerous and others less numerous in the gene pool." This is, of course, the familiar short, standard definition of evolution to be found in the synthesis (for examples, see Eldredge 1979). But with Dawkins the theme runs more deeply, as witness his views on the ontological status of biological entities usually considered to be implicated in the evolutionary process: while the genes are virtually immortal, in sexually reproducing species, the individual is too large and too temporary a genetic unit to qualify as a significant unit of natural selection. The group of individuals is an even larger unit. Genetically speaking, individuals and groups are like clouds in the sky or dust-storms in the desert. They are temporary aggregations or federations. They are not stable through evolutionary time. Populations may last a long while, but they are constantly blending with other populations and so losing their identity. They are also subject to evolutionary change from within. A population is not a discrete enough entity to be a unit of natural selection, not stable and unitary enough to be "selected" in preference to another population. (Dawkins 1976, p. 36).

Williams makes much the same point:

Low rates of endogenous change relative to selection coefficients are a necessary precondition for any effective selection. The necessary stability is the general rule for genes. While gene pools or evolutionary trajectories can persist little altered through a long period of extinction and replacement of populations, there is no indication that this is the general rule. Hence the effectiveness of group selection is open to question at the axiomatic level for almost any group of organisms. (Williams 1966, p. 114)

Dawkins, in his insistence (1976, p. 37) that "individuals are not stable things, they are fleeting" goes further perhaps than Williams in doubting the cohesive integrity of such entities as organisms, demes, and species (let alone taxa of higher categorical rank). His point is not that such entities are merely classes of constituent member entities; it is rather that such entities are too loosely bound together, or at any rate too ephemeral, to have the requisite stability for some appropriate selective process to "choose" between them. Thus, this particular point, hinging as it does on the ontological status of these entities, devolves to a matter of the packaging of genetic information. By the time Dawkins develops the argument to its extreme, organisms, demes, species, and so forth, as unstable packages of genetic information, become almost progressively, relatively unimportant items for consideration of the evolutionary process. The gene is all—by no means a novel view per se but one that is unusual and extreme in the intensity of its view that the gene is virtually the only entity of evolutionary interest in the biologic realm. Nor in his later *The Extended Phenotype* (1982) has Dawkins materially altered his positions on biological ontology or the levels of selection that are possible. But Dawkins's (1982) vision of a world of "replicators" and "vehicles" clarifies and extends both his ontology and his notions of the realms of possibilities vis-à-vis selection in a way that invites hierarchical interpretation (e.g., Salthe 1985), even though both Hull (1980) and Dawkins (1982) see little interest or benefit in an explicit adoption of hierarchy per se. Thus, I defer a discussion of some of Dawkins's later views to the next chapter.

Dobzhansky, of course, also saw information (he did not use that word) packaged in various sorts of ways within the organic realm. He openly wondered why there are such things as species and produced a theory to explain the problem. With Dawkins's thesis, the problem simply vanishes—or at least becomes irrelevant. I find that state of affairs, where large-scale packages of genetic information are not only subordinate to but absolutely irrelevant when considered alongside genes themselves, unfortunate. And, in what I construe as a milder position, Williams (1966, chapter 4) acknowledges that patterns of differential survival (at least, and possibly differential birth as well) of groups is "important" in evolution, but only in the sense of pruning, not creativity. Thus, as I shall argue in later chapters, a major component of the geometry of the history of life is effectively dismissed as trivial. Citation of college texts is a risky business, but I think the concurring words of Dobzhansky, Ayala, Stebbins, and Valentine (1977) on the subject summarize all that the core synthesis really has to say about selection above the level of pure natural selection. In their discussion of "group and kin selection," they write:

Related species compete for resources that both are in need of, and one species may outbreed and crowd out another. Thus, species introduced from foreign countries sometimes become pests or weeds, and eliminate or greatly reduce the abundance of native species. This is particularly striking on oceanic islands. Island endemic species often seem to have reduced competitive abilities, and they are given short shrift by introduced forms.

A more complex and interesting problem is whether group selection is also important intraspecifically. (Dobzhansky et al. 1979, pp. 125–26; italics added)

A scant three sentences on interspecific competition before moving on to the "more interesting problem" of intraspecific group selection suffices for this modern statement of evolutionary theory, and the ensuing discussion—which interestingly starts with Wright's model of semi-isolated colonies (this section of the book was written by Dobzhansky)—consists of two skeptical pages.

*Plasticity*

Williams (1966) also had much to say on the subject of the evolutionary plasticity of species. The subject is related to considerations of multilevel selection for two reasons. First, group selection pertains to group-level "fitness." Second, group selection benefits the group, even if it is neutral or deleterious to individual organisms. But Williams saw an additional problem in need of clarification. It had been (and remains) conventional among writers on evolution to speak of natural selection favoring the "survival of the species." On the latter point, Williams (1966) wasted few words; throughout his book he agrees with the usual conclusion that natural selection is very much ad hoc, adjusting populations to current environmental conditions. In no way can selection anticipate the future. Yet, after selection hones an adaptation, the chances of long-term survival of an entire species may reasonably be hypothesized to be enhanced; hence, selection is "for the good of the species." Williams makes clear that such long-term survival of the entire species is an entirely incidental byproduct of the efficacy of the adaptation in allowing organisms to survive and reproduce.

If natural selection has no eyes for the future, what then of the concept of plasticity, where there is a tradeoff between the value of narrowly focusing adaptations versus the retention of sufficient variation as a store for the modification of adaptations to meet the (inevitable) changes in environmental conditions? Williams writes:

There is no way in which a factor of necessity-for-survival could influence natural selection, as this process is usually formulated. Selection has nothing to do with what is necessary or unnecessary, or what is adequate or inadequate, for continued survival. It deals only with an immediate better-vs.-worse within a system of alternative, and therefore competing, entities. It will act to maximize the mean reproductive performance regardless of the effect on long-term population survival. It is not a mechanism that can anticipate possible extinction and take steps to avoid it. (Williams 1966, pp. 31–32)

The argument is thus symmetrical: biotic adaptation via group selection requires benefit to the group in contrast to neutral or deleterious consequences for the component organisms for its very detection—and Williams finds some, but little convincing, evidence for such in nature. Yet Williams readily imagines selection benefiting organisms over the short run (albeit through descendant populations), with long-term deleterious consequences for the group. And, as Eldredge and Cracraft have stressed (1980, chapter 6; see also Vrba 1980 and Eldredge 1984), there is more than anecdotal evidence suggesting that narrowly adapted (relatively stenotopic and/or anatomically, physiologically, and behaviorally "specialized" species) do in fact exhibit higher characteristic rates of extinction than their more generalized close relatives. What Williams and many other evolutionary biologists are saying is that the long-term consequences of within-population selection are side effects—or simply, in Williams's parlance, "effects" (though Williams himself did not extend the use of the term to this phenomenon; cf. Vrba 1980).

The point here is that most evolutionary theorists writing after the initial documents of Dobzhansky, Mayr, and Simpson have simply argued that "Dobzhansky's dilemma"—the tradeoff between the virtues and necessity of perfecting organic adaptations versus the need to retain sufficient variation for future evolution, is no problem at all, as selection simply hasn't eyes for the future. Logical indeed as this simple conclusion is, it allows core evolutionary theory (as opposed to ecological theory) to beg the basic question of why the organisms within some species have characteristically broad adaptations, while close relatives are much more narrowly niched (see Eldredge and Stanley 1984 for a number of examples). That the subject is broader than its casting strictly in terms of selection working on a moment-by-moment basis on a mere collection of sexually reproducing organisms begins to become clear again in the context of a hierarchically based evolutionary theory.

Thus, "levels of selection"—when resolved as it generally has been in favor of a reductionist, single-level point of view (and as internally logical as such discussions may be)—has had the the additional effect of removing some of the main issues that Dobzhansky (primarily) saw as crucial in the early days of the synthesis. Why do we have species? Why are some species (via their component organisms, to be sure) relatively broad-niched and others (closely related) relatively narrow-niched? I do not mean to suggest that these questions have been totally dropped in western intellectual thought; ecologists have always wondered about such things. But in terms of the formal structure and content of evolutionary theory, the turn downward and inward, to the genetics of populations and ultimately to the dynamics of genes themselves, at the very least took the attention of evolutionary theorists away from such matters—to the detriment, I firmly believe, of evolutionary theory.

## Variation

A fundamental cornerstone of the synthesis is that variation reduces to— because it arises at—the level of the individual genome, in the replication mistakes that we call mutations. Dobzhansky's hierarchical view has been retained in sufficiently clear form that all agree that this variation by mutation is injected into the population (deme) level, where it is subjected to a regime of natural selection (and the vagaries of drift) which has no reference to the mutation events themselves. Thus, in a general way, the nature of the origin of variation within populations, as well as the modes of determining its fate, seem clear and little changed from the early days of the synthesis.

But variation stands at the center of some problems that have arisen in recent years in evolutionary thought. In a way, as Lewontin has pointed out, variation can be (and has been) thought of as the flip side of selection; selection is an antivariation culling device. As Lewontin (1974, p. 195) characterized this view, "many are called, few are chosen." Indeed, the entire controversy over evolutionary plasticity (see above) is every bit as much a wrangle over why there is variation as over why selection hasn't removed it. Variation is to selection as ham is to eggs, though selection formally requires variation whereas variation exists independently of selection. Or does it? Lewontin claims (1974, p. 195) that the view that selection and variation are necessarily opposed is characteristic of those who adhere to classical Darwinian theory, "direct inheritors of the pre-Mendelian tradition." But, as we have seen, Dobzhansky himself (1937a) looked upon the relationship between variation and natural selection in just this way. Random, Wrightian processes, on the other hand, could in some cases promote variation (as when a species is divided into numerous semi-isolated colonies) and in others oppose variation.

There has been confusion over whether or not selection can be a cause of variation. Williams (1966, pp. 111, 112) cites Wright's (1945) initial discussion of group selection, in which he envisaged the spread of genes by drift (even when opposed by genic selection within small demes). The genes may be supposed to be favorable to the deme as a whole; demes thus increase in size and send out colonists to other demes. The genes spread despite being selected against within demes. "According to this theory, selection not only can act on preexisting variation, but also can help to produce the variation on which it acts, by repeatedly introducing the favored gene into different populations" (Williams 1966, p. 112). Williams then goes on to cite Simpson's doubts about the realistic nature of Wright's picture. Suffice it to say that no major block of theory has arisen which sees selection creating or even maintaining large pools of genetic variation for its (i.e., selection's) future use. Dobzhansky saw the advantages to a species of maintaining such a store of variability but never suggested that selection could actually act to produce or maintain it—an interesting situation inasmuch as Dobzhansky saw *genetic* advantages for the biotic world to be broken up into discrete species (to avoid inharmonious gene combinations) and concluded thereby that selection must be creating the discontinuities between species for that very purpose, a conclusion not generally adopted by the synthesis as a whole.

Lewontin (e.g., 1974), however, freely talks of selection contributing to or maintaining variation. Lewontin points out that the old "classical" school only saw the antagonistic relationship between variation and selection; yet balanced polymorphism and a number of other phenomena clearly show that selection can promote variation. We must only be careful with our words. Variation per se is not being selected because variation somehow conveys an advantage; Lewontin is speaking of traditionally construed genic (interorganismic) selection. In Williams's terminology, maintenance of variation is merely an effect of selection acting in certain ways, as in the selection of heterozygotes, such as in sickle-cell anemia.

The issue of variation is one of the strongest suits in the claim of Stebbins and Ayala (1981), cited earlier in this book, that all developments in evolutionary biology to appear since the synthesis was formulated can comfortably be accommodated within the synthesis. Perhaps the first formidable, overt challenge to the synthesis was essentially a theory of variation, or rather an extreme position taken on the relationship between selection and variation based upon a previously unsuspected generalization about the occurrence of variation in nature. Electrophoresis (Lewontin and Hubby 1966 and, of course, many others since then) began to reveal an astonishing amount of allelic variation in natural populations— variation that seemed inexplicable given the view of selection as being on constant watch, always culling the very best for the next generation. Indeed, the picture of a delicate balance between narrowly channeled perfection and the retention of *some* variation (fortuitous but potentially useful for further evolution) was exploded by the rampant variation documented from the late 1960s to the present day. Loci that code for enzymes turn out to be in some cases wildly polymorphic.

Clearly, such polymorphic loci create variant versions of enzymes that function about equally well, near enough so that selection cannot tell them apart. Their effects on relative fitness, in other words, are apparently minimal. Thus, the "theory of neutrality" (Kimura 1968), or "non-Darwinian evolution," was born. We have already seen that Mayr (1942) especially was accustomed to thinking of "selectively neutral traits," and Lewontin's statement on neutrality puts the empirical discovery of all this variation within a historical perspective of thought on the issue of variation and shows at least one geneticist's way of accommodating the new observations within a familiar theoretical matrix, as Stebbins and Ayala (1981) claim is generally possible with such developments. Under the "neoclassical theory," Lewontin says:

Thus the so-called neutral mutation theory is, in reality, the classical Darwin-Muller hypothesis about population structure and evolution, brought up-to-date. It asserts that when natural selection occurs it is almost always purifying, but that there is a class of subliminal mutations which are irrelevant to adaptation and natural selection. This latter class, predictable from molecular genetics and enzymology, is what is observed, they claim, when the tools of electrophoresis and immunology are applied to individual and species differences. (Lewontin 1974, p. 198)

Variation, so crucial to Darwin's argument (as Lewontin justly claims), has remained of vital concern to the synthesis. The consensus view is certainly that variation arises ultimately in the genome through a number of deterministic causes. Such mutations are random only with respect to the needs of an organism. Another way of saying this is that mutations are random with respect to the regime of natural selection. Sexual reproduction, via recombination, inadvertently scrambles up variation, presenting new genetic combinations to selection, and incidentally providing a store of variation for future evolution. Thus, maintenance of variation and its distribution are mere effects of mutation, sexual reproduction, selection and drift, and the structure of populations within species. Variation as a store for future evolution is not the product of natural selection, adaptive though it may seem to be. This is the prevailing and unchanged thesis on variation held by the synthesis, and most of these points, again, seem inarguable.

But is that all that might be said about variation and its significance to evolutionary theory? To consider variation within populations is to consider the structure of information among organisms within populations. Dobzhansky was concerned with species as packages of such genetic information, recognizing (at least in some passages) among-demic, within-species "packages"—a view we owe, of course, to Sewall Wright and a view only now beginning to receive the serious attention I believe it deserves.

Hierarchy theory generalizes the entire approach to the distribution of variation in nature. It takes the position that just because novelty (and noise) in information arises in the genome is no reason to suppose that its only significance for the evolutionary process resides at that level, or at the next higher level (within populations, where the bulk of the synthesis still tends to see the significance of variation residing). Of course, it will always be important to consider variation at those levels, but not to the exclusion of higher and lower levels. Indeed, the impact of a hierarchical approach to variation has some ironic consequences for Mayr's (1942, and many later references) scathing attacks on typology, as I shall discuss in chapter 7.

## THE SYNTHESIS: A DISTILLATION

What, then, *is* the synthesis? What does it say? Genes are probably foremost among the biological entities typically alleged to participate in the evolutionary process. In this context, Williams's (1966, p. 24) definition and characterization of the gene as "that which segregates and recombines with appreciable frequency" is all that is needed for conventional evolutionary theory. Genes are atomistic, particulate, historical entities, their historicity never better appreciated than in Dawkins's (based on Williams's) vision of genes as potentially immortal. The synthesis, too, sees organisms existing in the real world, their evolutionary significance riding in their physical (phenotypic) manifestation of the genome, a requisite for the interplay between environment and genome via natural selection. Similarly, populations (demes, isolated colonies[4]) may exist but are ephemeral. As both Eldredge and Salthe (1984) and Vrba and Eldredge (1984) have pointed out, demes figure importantly in the synthesis directly in proportion to the degree that Wright's work (1931, 1932) is seen as integrated with the synthesis. Thus, Dobzhansky (1937a) insists that species are commonly broken up into small, semi-isolated colonies, and he utilizes Wright's work as a cornerstone of his general picture of the evolutionary process. But the synthesis as a whole has tended to treat Wright's demes, indeed his entire "shifting balance theory," more as a special case than as a central thesis.

I have reviewed the ontological status of species in the synthesis in detail above. In brief, species have generally been seen to be somewhat less ephemeral than organisms and demes, though by the time we get to the hyper-reductive views of Dawkins, species seem to have faded from evolutionary consideration nearly altogether. In the synthesis, on the whole, species are more or less concrete entities at any one point in time; through time they take on the properties of classes, not necessarily being formally temporally bounded, and have no special further significance. This characterization is largely true despite early statements (e.g., Dobzhansky 1937a) on the supposed adaptive *purposeness* of species and later positions of some of the architects of the synthesis (e.g., Mayr 1963) on the roles species themselves may play in the evolutionary process. The synthesis tends to see monophyletic taxa (when considered at all) as, at most, genealogically interconnected strings of species. But taxa of higher categorical rank need not be construed as monophyletic, as I have argued in connection with the explicit adoption of "grades" in evolutionary analysis. Indeed, the synthesis has always tended to treat macroevolution as a problem of *large-scale* transformation of phenotypic characters; hence "taxa of higher categorical rank" are of significance predominantly, almost wholly, as *name-bearers of collections of phenotypic features* of varying magnitudes of distribution.

This, then, is the cast of characters in the discourse of the synthesis: genes, organisms, populations, species, and taxa of higher categorical rank. Only the first two are unequivocally seen as "hard" historical entities. Otherwise, species are crucial, but their ontological status depends very much upon where one is in the development of the argument. The ontological status of the other pieces of Bunge's (1977) "furniture" is even more wispy. Regarding subdivisions of the genes, nothing is said because nothing was known in the early days of the synthesis, and little until recently about the molecular structure of the genome seemed to call for anything other than a Williams-like definition of the gene in an evolutionary context. Of the sorts of ecological entities—communities, ecosystems, and so forth—nothing is formally or explicitly stated in the synthesis.

As for process, the synthesis quite naturally restricts itself to those that affect the origin, development, maintenance, and termination of the entities it sees as evolutionary participants. Mutation, recombination, selection–adaptation, drift, and speciation are paramount. The nuances here are numerous, as anyone even casually familiar with population biology–genetics well knows. The interactions between the processes and such variables as population size, number of loci in the model, and so on, continue to provide perhaps the bulk of the theoretical literature on evolutionary processes. Missing from the synthesis to date, however, are possibly relevant molecular processes (for the simple reasons given above) and formal and serious considerations of processes affecting birth, death, and sorting (Vrba and Eldredge 1984) of populations, species, and taxa of higher rank. Also missing, interestingly, is any formal integration of (1) developmental processes and (2) ecological processes—exceptions being formal considerations of the sources of selection forces in nature (e.g., Bock 1979). But ecological processes, and especially large-scale ecological entities, are by no means truly integrated into the "synthesis."

Thus—to distill a distillation—I find Mayr's (1980a, p. 1) brief encapsulation of the synthesis surprisingly good given its extreme brevity; the synthesis really does come down to a simple statement that "gradual evolution can be explained in terms of small genetic changes ('mutations') and recombination, and the ordering of this genetic variation by natural selection; and the observed evolutionary phenomena, particularly macroevolutionary processes and speciation, can be explained in a manner that is consistent with the known genetic mechanisms." One can adopt this summary statement without doing profound violence to that great and somewhat diffuse body of work that is the modern synthesis. I would only add that "is consistent with" can just as well be read as "(the observed evolutionary phenomena) are effectively caused by"—and one still has an accurate, pithy summary of the synthesis. The synthesis today really does seem to be the explanation of all manner of evolutionary phenomena through invocation of the simple mechanics of among-organism, within-population genetic change. It is far less the melding and unification of a set of scientific disciplines than it is the explanation of the data of a number of disciplines by the "statics and dynamics" of genetics. Only if one defines the synthesis as "all evolutionary biology" can one agree with Stebbins and Ayala (1981) that the umbrella of the synthesis is sufficiently broad to embrace all the empirical and theoretical wrinkles that have appeared since its inception.

So far, I have examined only those newer themes that the synthesis does seem to have handled fairly well—issues that have been accommodated by simple expansion of earlier themes that were present in the synthesis all along. That the bulk of synthesis-inspired theory and research in evolutionary biology in the 1980s broaches no expanded list of the kinds of entities (and processes) seen to figure in the evolutionary arena (expanded, that is, over the list of entities and processes that have always been considered in evolutionary theory) is best shown through direct examination of such an expanded list, a task I begin in the following chapter and pursue throughout the remainder of this book. It is certainly true, however, that a thorough review of current evolutionary biology would reveal attention to a host of subject areas not directly considered in the early documents of the synthesis, including sociobiology (and group and kin selection) and coevolution, plus explicit attention to a wide range of particular problems in both theoretical and "practical" population genetics (certainly including most optimality arguments in the analysis of the origins and adaptive nature of particular behaviors and structures through natural selection). The field is vital—as far as it goes. But its vision—in terms of problems, processes, and the very range of entities that it sees taking part in the evolutionary process—remains limited. With the exception of group selection, evolutionary biology remains steadfastly focused on what Williams calls organismic—as opposed to biotic—problems, entities, and processes. In the next three chapters I shall simply urge an explicit, formal *addition* of biotic considerations to the traditional organism-centered focus of the modern synthesis. I shall begin with a discussion of recent themes that, on the face of it, the synthesis has not handled adroitly.

## NOTES

1. Huxley acknowleged that grades and clades may and often *do* coincide; thus, Mammalia may be construed as a monophyletic grade. But parallelism and convergence, according to Huxley, impede recognition of monophyletic groups and, instead, create gradal groups. Hence, grades do not need to be monophyletic; they often are polyphyletic—yet no less real to Huxley.

2. It is probably because debates about species selection appear to have triggered many of the discussions of hierarchy that the confusion that sees multilevel selection as synonymous with hierarchical evolutionary theory has crept in.

3. A word of clarification on Williams's terminology in this passage: *biotic* means any naturally occurring collection of individual organisms, such as a population, deme, or species. *Organic* refers strictly to aspects of organisms. Genic selection produces organic adaptations; group selection produces biotic adaptations.

4. The terms *demes, populations,* and *colonies,* in the context of the synthesis, generally are synonyms. As mentioned already, I will use *population* only in an ecological sense and *demes* only in a genealogical (genetic) informational sense in later chapters.

# 5

# Toward Hierarchy: Trends and Tensions in Evolutionary Theory

"Evolutionary biologists are currently confronted by a . . . dilemma: If they insist on formulating evolutionary theory in terms of commonsense entities, the resulting laws are likely to remain extremely variable and complicated; if they want simple laws, equally applicable to all entities of a particular sort, they must abandon their traditional ontology. This reconceptualization of the evolutionary process is certainly counter-intuitive; its only justification is the increased scope, consistency, and power of the theory that results." (Hull 1980, pp. 316–17)

Is the synthesis a complete and satisfactory evolutionary theory? It would be a trivial and derogatory exercise indeed to depict the synthesis in utterly simplistic terms and then turn around and conclude that it is incomplete and unsatisfactory. Though I have subjected the synthesis to a series of purifying distillations through the course of this book so far, I have in mind by no means solely Mayr's (1980, p. 1) two-sentence summary of the synthesis as I pose the question of just how complete, workable, and satisfactory a theory of the evolutionary process it is. No serious student of evolutionary theory could ever claim that the modern synthesis is "just population genetics." Many more phenomena are included than the statics and dynamics of genetic change in populations.

What *does* seem to be true of the synthesis in general is that it focuses its concerns on a certain range of biological entities and attendant processes, and espouses attitudes and ontological positions on others that appear (to me) to exclude the latter from effective integration into the theory per se. Specifically, the synthesis focuses on genes; their replication, recombination, and mutation; and the fate of allelic variation within populations. But species—certainly not the sole province of population genetics—are very much a part of the synthesis, if equivocally so. If it is true that most evolutionary phenomena considered by the synthesis are construed, at least in principle, to be explicable in terms of the dynamics of selection and drift of allelic variation in populations, it is not because other sorts of phenomena, such as macroevolutionary trends, are alleged not to exist. The synthesis takes the (on the whole commendable) attitude of the Missourian who must be shown. We have a highly corroborated theory of the origin, maintenance, and modification of adaptations— through pure, narrowly defined natural selection. The burden of demonstration lies on anyone who would maintain either that some other process builds such (organismic) adaptation or that an additional process (or more than one) is also at work in evolution. Certainly the entire discussion on levels of selection, a discussion to which I return in this chapter, is structured in this general sort of way.

Thus, I am aware that the synthesis is not a monolithic and utterly coherent statement. I know that it addresses many things. I know, too, that when I characterize the ontological position of the synthesis on, say, monophyletic taxa as being essentially classlike, quotes are available to show that some authors, including certainly the very architects of the synthesis (even Simpson), sometimes saw such entities as individuals. When I claim, as I shall, that such entities as communities are not effectively integrated into the synthesis—indeed, the entire relation of ecology to evolutionary theory remains almost bizarrely nebulous (but see Ghiselin 1974b for an interesting attempt)—many who teach an amalgam of both ecology and evolution at universities will bridle. Remember that Dobzhansky was incensed in 1937 as he dismissed the notion that genetics pays no heed to ecology. Yet I will conclude that there is no formal, effective integration of ecology with the entities *usually* construed as evolutionary—entities that form the "genealogical hierarchy" (Eldredge and Salthe 1984; Vrba and Eldredge 1984; chapter 6 herein).

There has been an undue amount of confusion in recent years about the proposals of some of us who see less than an utterly complete theory in the synthesis (e.g., Gould 1980a; 1982b). In particular, it is often perceived that a hierarchically based evolutionary theory—a theory that sees ontologically based hierarchies of evolutionary entities—constitutes a strict and total alternative to the synthesis. For example, Thompson (1983, p. 450) writes that "Gould, Stanley and Eldredge [must] provide an alternative theory of heredity" to avoid having patterns explained merely by the accumulation of the effects of selection within populations. According to Thompson (1983, p. 450), we are producing an "irreducible evolutionary hierarchy."

Such egregious misunderstanding must be avoided if the rest of this book is to be read and understood. I—along with Gould, Vrba, Salthe, Damuth, Stanley, and all others who have been moving in the general direction of hierarchy theory—have no intention of junking the synthesis (were such an action possible). Specifically, no one denies that there is design in nature—design called adaptation.[1] Adaptations in this conventional sense are attributes of organisms—anatomical, behavioral, physiological phenotypic features. It is widely acknowledged that there is some control of the phenotype by the genotype; even if the particulars of the genetic basis of such features are not understood for any given instance, the bill structure of pileated woodpeckers and the necks of giraffes are inherited with astonishing fidelity. We have a theory—natural selection—which is highly corroborated in field, experimental, and mathematical inductive examination. That theory provides a deterministic but statistical mechanism to alter phenotypes via modification of the underlying genotypes. Tautological problems aside, removing natural selection as a force in evolutionary theory is as easy—and as desirable—as taking the gin out of a martini.[2]

No, the problem with contemporary evolutionary theory is not that its essential neo-Darwinian paradigm is incorrect. The problem is that the consistency argument of the synthesis (as developed, for example, in Mayr's (1980a, p. 1) statement; see also Gould 1980a), is itself troubled. That argument says that the core neo-Darwinian paradigm (the theory that deals with the origin, maintenance, and modification of within-population genetic structure) is *consistent* with all other known evolutionary phenomena. This credo, innocuous and undeniable as it is, has been expanded to mean that the neo-Darwinian paradigm of selection plus drift, are both necessary and sufficient to explain all other known evolutionary phenomena. My position here, and the position of all other doubters of the completeness of the synthesis that I know of, is simply that the neo-Darwinian paradigm is indeed necessary—but is not sufficient—to handle the totality of known evolutionary phenomena. And it may not even be necessary to explain certain particular phenomena. It is thus not a matter of either/or. The first part of Mayr's one-two-punch statement— a shorthand version of the neo-Darwinian paradigm—must remain intact, at least until an evolutionary biologist with data and theory directly relevant to the within-population level of allelic variation falsifies it or otherwise casts grave doubts on its validity. This is unlikely to happen, yet not a totally impossible or unimaginable occurrence.

It is the second part of Mayr's statement—the "extrapolationist" segment—which alone is under challenge. It simply seems to be insufficient for two different, albeit related, reasons. The first is strictly epistemological: there is no way directly to test hypotheses of evolutionary process with any data other than those expressed in terms of gene frequencies. This leaves the majority of biologists out and vexes at least some of them. They would like to see theories of the evolutionary process couched in terms directly relevant to the phenomena they study. The second reason for questioning, not the *necessity* but the *sufficiency* of the neo-Darwinian paradigm is more complexly subtle yet far more important withal. It is *ontological*: in order to use the neo-Darwinian paradigm to explain everything, we must stress the existence of some biological entities while ignoring (even denying) the existence of others. My position here is that codons and monophyletic taxa are every bit as much coherent historical entities—individuals—as are organisms, and they simply must be regarded explicitly as such in evolutionary theory.

Moreover—and here is the central contention of this book—each of these sorts of individuals are nested in the biological realm: instances (individuals) of one level form the subunit parts of individuals at the next higher level. It is this natural hierarchy (there are actually five such hierarchies relevant to the evolutionary process—see chapters 6 and 7 herein) that dictates the necessity of adopting a hierarchical structure of evolutionary theory, at least if that theory is to embrace all evolutionarily relevant biological entities. It is not the demonstration, or any concomitant necessity, of hierarchically arrayed processes that forces us to consider a hierarchically structured evolutionary theory. Such matters of process remain contentious and highly debatable. There is no ineluctable demonstration, for example, of any form of group selection as a general phenomenon. Rather, it is the existence, seemingly undeniable, of hierarchical arrays of genealogical and ecological entities, all generally conceded to be somehow the product of and thereby relevant to the evolutionary process, that forces us to adopt a hierarchical approach. Thus the fundamental issue is indeed ontology, a matter I pick up and pursue in chapter 6. For the remainder of the present chapter, I pursue an amalgam of epistemological and ontological considerations which either foster doubt on the full ability of the synthesis to explain how the evolutionary process actually works or directly lead to considerations of hierarchy in evolutionary theory. I shall then turn to an explicit consideration of hierarchy. I will urge a return to, plus clarification and expansion of, the avowedly hierarchical scheme that Dobzhansky advocated in 1937.

## BASIC EPISTEMOLOGICAL PROBLEMS

One class of objections to the synthesis is simply put: of the gamut of the actual historical events in evolution, ranging from particular mutations to the extinction of an entire order or the development of particular communities, only a small subset are explicitly addressed by the synthesis. Classes of historical events form the recurrent patterns that the theory addresses, yet the theory speaks directly only to the issues of mutational origin, plus maintenance and modification of genetic variation within demes. Speciation is included as something of a codicil to the main argument. In any case, the relationship of speciation to macroevolution remains vague. In general, the synthesis has ignored speciation when it has confronted the larger-scale phenomena of macroevolution, preferring to see such patterns as trends, adaptive radiations, and the like as merely a wholesale accumulation of conventional Darwinian adaptive change (see Eldredge and Cracraft 1980 for a more extensive discussion of such macroevolutionary theory). What we have, then, are two ships passing in the night. The hold of one is crammed with phenomena either ignored (ecology, developmental biology) or only vaguely addressed (species, monophyletic groups, the molecular anatomy of the gene), while the other ship bears an explanatory theory only alleged to be relevant to such phenomena. In the popular parlance of contemporary philosophy of science (as seen at least by some scientists), such a situation renders much of evolutionary theory *untestable*. There is simply no way to evaluate a statement about fossils that is written in the language of genetics.

These points are not new, nor will I dwell on them long. As I have remarked elsewhere (Eldredge 1982b), the methodological crunch, where a theory of the genetics of populations is asserted to account for many if not all such evolutionary patterns, merely makes matters difficult for those of us who are not population geneticists. It does not thereby invalidate the theory. The synthesis may be correct in all its particulars and stand as a complete evolutionary theory as well, and our difficulties in applying the theory rigorously to data sets that involve something other than generation-by-generation change in gene frequencies might merely be the bad breaks in the game. Should Carson (1981) be correct in stating that the kinetics of evolution do in fact lie strictly within the province of genetics, it would indeed follow that the epistemological difficulties that we might have in testing that theory with our own sets of nongenetic data would amount to nothing more than inconvenience plus a dash of disappointment.

But Carson is probably not correct in his bald assertion that so eloquently captures the essence of the synthetic view. He is probably wrong, not because of the aforementioned epistemological bind, nor because we have hard and fast manifest evidence of a slew of evolutionary processes amounting to a multiscale evolutionary kinetics of evolutionary change; the big problem with the synthesis in general and Carson's remarks in particular is the mistaken or at the very least only partially correct and complete ontology espoused by the synthesis. It is the empirical and theoretical work that has clarified—even radically altered—this ontology that more than anything else forces us to admit the inadequacies of the synthesis. We are led ineluctably, I think, to see in its stead a hierarchical structure for the sorts of entities that are the players in the evolutionary game. And that ontologically based hierarchical structure in turn forces us to think about the evolutionary process in explicitly hierarchical terms.

## MOLECULAR BIOLOGY

It is certainly true that the founding fathers of the synthesis could hardly be faulted for failing to anticipate recent advances in our knowledge about the gene. But it is equally true that treatment of physiological genetics as a black box made several assumptions, including the sanctity of the Weismann doctrine (that the germ line affects and effects the soma unidirectionally; there is no feedback of somatic change to the germ line). It was further assumed that, because genes are atomistic particles strung out along chromosomes, apart from rearrangements within and among homologous chromosomes, there would be no mechanism other than natural selection and genetic drift that could act systematically to alter allelic frequencies within populations. Molecular biologists have recently been scrutinizing these fundamental black-box assumptions.

The Weismann doctrine, that bastion of anti-Lamarckian argument, has recently received its strongest experimental attempt at falsification—and has escaped unscathed. Dawkins (1982) describes the whole affair as a "scare," as indeed the overthrow of Weismann's doctrine would throw most Darwinian-based evolutionary theory (including the present work) into disarray. The experiments of Steele (1979) and Gorczynski and Steele (1981), which led to the apparent inheritance of acquired chacteristics of the immune system (see Dawkins 1982 for a recent summary), simply have failed the test of replicability—no one has been able to get the same results. The collective sigh of relief within orthodox evolutionary biology was almost audible, for the second-line argument surely would have been, "Well, it's only one minor aspect of the immune system and doesn't pertain to the basic adaptations of morphology, physiology, and behavior that form the standard phenotype." And that is wishy-washy.

So much for the scare. But molecular biology has revealed a complex structure for nuclear, mitochondrial, and chloroplast DNA and the various forms of RNA, a complexity that is apparently related to the functional control of these elements. The structure is strongly hierarchic. Some aspects of genomic organization seem at least to some molecular biologists to violate the supposition that only selection and drift can disrupt Hardy-Weinberg equilibrium. Primary among these observations, of course, is the existence of transposable elements, which apparently come in a variety of classes. Families of repeated sequences of noncoding DNA have intrigued a number of biologists, who have formulated a number of variant explanations for such gene families.

Orgel and Crick (1980) and Doolittle and Sapienza (1980) think of the families of repetitive, noncoding DNA as "selfish" DNA. The limit to the takeover and spread of these families within the genome, according to them, comes when the repeated elements begin to interfere with the "normal" functions of the genome. There is nothing particularly unorthodox from an evolutionary point of view in the selfish DNA concept; indeed, Dawkins (1982) has written about selfish DNA as a natural adjunct to his own notions of selfish genes, and though Dawkins speaks of the Necker cube (gestalt-switch) aspect of looking at evolution from the point of view of the organism or the genes it carries, he concedes that his is an argument of emphasis, not a qualitative departure from Darwinian orthodoxy.

Dover (e.g., 1982) and colleagues see these repeated patterns of noncoding DNA somewhat differently. Claiming that such families are typically strikingly homogeneous *within* but not *among* species, Dover hypothesizes that genic elements (both transposable and not) can preferentially modify the frequencies of allelic forms. The genomes of all the organisms within a species can thereby become homozygous for these families by a mechanism (molecular drive) having nothing to do with natural selection or genetic drift.

My purpose here is not to tout any particular set of novel theories in molecular biology. Indeed, I confess barely to understand them. My sole purpose in addressing the subject is to point out that there are biologists working with phenomena *other* than species and monophyletic groups who are experiencing ontological and epistemological difficulties in explaining their data with the tools afforded by the synthesis. Difficulties with the synthesis come from lower as well as higher levels of biological phenomena. That this is true is plain enough (though the validity of any particular special claim is, of course, not thereby established), and the larger point, that we therefore require a hierarchically structured evolutionary theory, is also not automatically established. But the current ferment in molecular biology certainly suggests that problems with the synthesis derive from several different levels of biological organization.

## DEVELOPMENTAL BIOLOGY

"Ontogeny recapitulates phylogeny." All biology students have been taught that maxim, then forced to unlearn it as hokum, only to see it resurrected as a useful guide to unraveling the complex history of life. Gould (1977) has given us a comprehensive review of the nexus of ontogeny and phylogeny, mainly from a process point of view. There is no doubt that the development of organisms has much to tell us about the nature and sequence of the phylogenetic acquisition of evolutionary novelties, and therein lies the interest of systematists and paleontologists in ontogeny, for the most part (e.g., Eldredge and Cracraft 1980, p. 58).

But what of the nexus between the process of development and the processes of evolution? The somatic hierarchy of organisms is the most obvious of all biological hierarchies. Diversification and transformation of the fertilized egg into the adult organism, still biology's outstanding mystery, must surely have something to do with evolution, if only because it is change in the somas of organisms that has always been taken as the prime manifestation of evolutionary change.

Yet developmental biology has imparted to evolutionary theory no lasting notions of process. Acceleration, neoteny, and the like sound like processes but in reality are descriptions of patterns of modified relative timing of developmental events. Developmental processes may affect the direction of phenotypic evolutionary change, through various constraints that center on the issue of just how far developmental pathways may be disrupted or modified without sacrificing embryonic viability. But are there processes of development that directly govern patterns of change through time in a more positive, "creative" fashion?

A recent trend in developmental biology—seen, for example, in the works of Goodwin (e.g., 1982), Ho and Saunders (1979), and Oster and Alberch (1982)—stresses the relative independence of the developing system from the underlying genome (see Charlesworth et al. 1982 for an opposing view). Whether preformationist or epigeneticist, evolutionary theory traditionally assumes that the development of the phenotype closely reflects instructions from the underlying genotype. The movement away from this view surely has evolutionary implications. Most attacks on the Weismann doctrine, after all, question the sanctity of the view that the germ line is free from somatic influence. Questioning the pervasive control of the developing soma from the ultimate germ-line source, the zygote, is something else again.

I will conclude in the next chapter that the somatic hierarchy, which indeed is unfolded each time an organism undergoes development, is a "special case" hierarchy of no profound significance for evolutionary theory, meaning ideas about how evolution works. Yet, if development is less clearly linked to the genome than tradition would have it, there are basic implications for the grand partnership of selection modifying gene frequencies via the phenotype. *That* central dogma, it seems certain, is practically inviolate and likely to remain so. Still, we need a more formal structure for a better integration of developmental processes with the mainstream of biological thought. And that structure is provided by the genealogical hierarchy.

## THE ONTOLOGY OF IT ALL

Hull (1980), in the quotation that stands at the head of this chapter, has said that there will be no advance in evolutionary theory, no recognition of "simple laws," unless we change our ontology from a recognition merely of "commonsense" entities to one that sees those *other* "things"—populations, species, monophyletic taxa, ecosystems—as every bit as much historical entities as are genes and organisms. And Ghiselin (1974a) entitled his seminal paper on the subject "A Radical Solution to the Species Problem."[3]

What is this radical shift in ontology? It is deceptively simple: common sense, as Hull says, tells us that organisms exist. They are spatiotemporally bounded entities, with beginnings, histories, and endings. They are the paradigmatic individual—to the extent that we often use *individual* to mean *organism*. Likewise, we treat genes as historical entities—from the beads on the chromosomal strand to the complex arrangements of base pairs into the variety of functional (and noncoding if not nonfunctional) parts of the modern conceptualization of the gene.

It was Ghiselin's (especially 1974a, but cf. his citation to Ghiselin 1966 and Hull 1974) singular contribution to argue concisely that species are not unbounded classes but are rather to be construed as individuals. Ghiselin (1974a, pp. 536, 542) freely acknowledged the long history of the notion. I have already (chapters. 2–4) looked at the ontological status of species in the synthesis in detail. In a nutshell, the synthesis treats species not as purely unbounded classes, but as clearcut "real" individuals when sympatric species are considered, epistemologically problematic but no less "real" individuals when allopatric species are considered (i.e., closely related species), and as classlike entities with only arbitrarily definable temporal boundaries when their existence through time is explicitly broached. (This brief encapsulation is based chiefly on Mayr 1942; see chapters 3 and 4 herein.) The ontological status of species in the synthesis very much depends upon context.

Ghiselin (1974a) sought to change all that, and I explore the ontological status of species in detail in the next chapter. It is important here to analyze Ghiselin's and Hull's contributions, as they directly anticipate recent developments in hierarchy theory. Ghiselin's remarks on biological entities *other* than species are important in this respect:

> The perplexing issues of the defining properties and ontological status of the categories lower than the species may be clarified in a similar vein. Subspecies are like local branches of a widespread industrial enterprise, but the degree of differentiation can hardly be so distinct as it is at the species level, for there can be all sorts and degrees of closure, and closure is not the same as differentiation. Nonetheless, subspecies may be expected to each have a characteristic local mode of competition, and what has been called their "reality" may be compared to that of any local productive unit. Yet as I have pointed out (Ghiselin 1966) some lower categories do not refer to populations: the morph, the *forma sexualis*, the monstrosity, etc. These are strictly nominal classes, and for that reason are best left out of the hierarchy.
>
> Are the taxa ranked at categorical levels higher than the species merely conceptual? The economic analogy here seems rather remote. Is the automobile industry real? One's answer depends upon his metaphysics, but at least industries have not the same ontological status as firms. Whatever answer is best, it may be worthwhile to explore the possibility of treating higher taxa as sectors of the natural economy. (Ghiselin 1974a, p. 540)

Clearly, Ghiselin thought of the possibility that clarification of the ontological status of species as historical entities, as individuals, could be logically extended to embrace other sorts of biological entities. He also clearly saw that these entities form a hierarchy, though he didn't elaborate on its nature.

Hull has pursued the notion of the individuality of species (e.g., Hull 1976, 1978, 1980) in detail. Hull, too, has explored the possibility that all manner of biological entities may also be construed as individuals, though he remarks (1980, p. 311) that "two assumptions are at fault" for the stalemates he believes evolutionary biologists have reached in their analysis of "several problems." The problems he cites (p. 311) are "the presence of so much genetic heterogeneity in natural populations, the prevalence of sexual forms of reproduction in the face of an apparent 50% cost of meiosis, and the difficulty of explaining how selection can operate at higher levels of organization"—problems already encountered in this book which virtually demand a hierarchical approach to their solution. The two faulty assumptions that Hull sees (p. 311) are fascinating: "(a) the view that genes and organisms are 'individuals' while populations and species are 'classes' and (b) our traditional way of organizing phenomena into a hierarchy of genes, cells, organisms, kinship groups, populations, species, and ecosystems or communities." I agree with Hull entirely on his first point. I cannot follow him on his second; reorganization of our ontology of such biological "things" seems automatically to imply hierarchy, and there is little evidence that hierarchy has been given the attention it seems to deserve. While we need not follow the exact form of the hierarchy in Hull's list, nonetheless recognition of the two process hierarchies (i.e., the ecological and the genealogical; see chapter 6 herein) follows automatically from correcting the first of Hull's two faulty assumptions: *all* of those entities (and more not listed) are individuals, not classes. I return to Hull's (and Dawkins's) work shortly in connection with a hierarchically minded discussion of levels of evolutionary processes, particularly selection. It is difficult to keep ontological concerns per se out of discussions of selection; the feeling one gets from the recent literature is that such ontological considerations—indeed, considerations of hierarchy—are only (or perhaps most) interesting from the standpoint of selection (see Gould 1982b on the "strong" argument for hierarchy; Dawkins 1982 seems to feel the same way for different reasons). Yet, as I develop in detail in the next chapter, it is purely ontology that forces us, just as it did Dobzhansky, to recognize a hierarchically structured biota and thus a theory of evolution explicitly addressed to hierarchy.

What are the formal consequences of seeing species, communities, monophyletic groups, chromosomes, and so forth as spatiotemporally bounded historical entities? There is an immediate, necessary epistemological consequence: we not only can, but we *must* frame theories of process that explicitly address all such identified members of the biological realm. Using species as the paradigmatic example, I cite Dobzhansky's graphic reduction of the history of life as a point of departure:

The enormous diversity of organisms may be envisaged as correlated with the immense variety of environments and of ecological niches which exist on earth. But the variety of ecological niches is not only immense, it is also discontinuous. One species of insect may feed on, for example, oak leaves, and another species on pine needles; an insect that would require food intermediate between oak and pine would probably starve to death. Hence, the living world is not a formless mass of randomly combining genes and traits [ever faithful to the first two editions of *Genetics and the Origin of Species!*], but a great array of families of related gene combinations, which are clustered on a large but finite number of adaptive peaks. Each living species may be thought of as occupying one of the available peaks in the field of gene combinations. The adaptive valleys are deserted and empty.

Furthermore, the adaptive peaks and valleys are not interspersed at random. Adjacent adaptive peaks are arranged in groups, which may be likened to mountain ranges in which the separate pinnacles are divided by relatively shallow notches. Thus, the ecological niche occupied by the species "lion" is relatively much closer to those occupied by tiger, puma, and leopard than to those occupied by wolf, coyote, and jackal. The feline adaptive peaks form a group different from the group of canine "peaks." But the feline, canine, ursine, musteline, and certain other groups of peaks form together the adaptive "range" of carnivores, which is separated by deep adaptive valleys from the "ranges" of rodents, bats, ungulates, primates, and others. In turn, these "ranges" are again members of the adaptive system of mammals, which are ecologically and biologically segregated, as a group, from the adaptive systems of birds, reptiles, etc. The hierarchic nature of the biological classification reflects the objectively ascertainable discontinuity of adaptive niches, in other words the discontinuity of ways and means by which organisms that inhabit the world derive their livelihood from the environment. (Dobzhansky 1951, pp. 9–10)

Species and monophyletic groups are treated in this passage as (temporally) unbounded groups. The *real* items of interest are the suites of adaptive characteristics of organisms that change through time. Stages of that transformation—plus the morphological gaps that appear between adaptive peaks—give the properties of the classes: the canids, felids, and ursids, the carnivores, the mammals, and so forth. Ghiselin's characterization of the synthesis is particularly vividly borne out in Dobzhansky's passage. The point is that the explanation of the history of life in such an adaptive scenario is only possible if species and taxa of higher categorical rank are treated only as classes defined by the very same properties that are seen to be undergoing adaptive modification through natural selection. It is in this sense that various authors looking at macroevolution in recent years (e.g., Eldredge and Cracraft 1980; Gould 1980a) call the synthesis reductive. And such reduction is impossible if species are construed to be individuals.

Ghiselin (1974a, p. 538) sees species as "the most extensive units in the natural economy such that reproductive competition occurs among their parts." But a generalized version of the "biological species concept" suffices equally well: a species (i.e., among sexually reproducing organisms) is a "diagnosable cluster of individuals within which there is a parental pattern of ancestry and descent, beyond which there is not"—to cite one segment of the definition in Eldredge and Cracraft (1980, p. 92). Species are reproductively coherent communities. What makes them reproductively coherent entities is a certain subset of the phenotype (in its broadest sense) of its member organisms—those geared to the recognition (again, in the widest sense) of conspecific prospective mates. The rest of the adaptive features of the member organisms are irrelevant to the simple fact of existence of any species—its ontological status as a spatiotemporally bounded entity. These other adaptations, geared as they are for the procurement of energy, for making a living, (that is, adaptations in the usual sense of the term, and certainly in the sense Dobzhansky was using in the passage just cited), pertains to the position of organisms and populations within an utterly different (ecological) hierarchy. Organismic adaptations have effects on the relative survival, and even potential for speciation, of species. Thus, I differ with Ghiselin (1974a, 1974b), but only to the extent that he and his readers confuse the *analogy* between economics and such genealogical entities as species with some stronger form of equivalence between them; species seem not to be individual participants in the "natural economy," nor is reproduction to be confounded with the economic aspects of biological organization.

Nor must species have "emergent" properties simply to exist. Most attributes of species, such as eurytopy and stenotopy, are seen as the additive properties of component organisms. I shall consider other possibilities when I take up levels of selection once again below, but in general the search for such emergent properties converges on Williams's (1966) unsuccessful search for what he called biotic adaptations.

What all this amounts to in the present context is that species as reproductive communities, spatiotemporally bounded, cut right across the selectively (or otherwise) generated spectrum of phenotypic diversity. It's the old "diversity and discontinuity" double theme of Dobzhansky and Mayr all over again. Species are not mere breaks in the flow of some stream of anatomical continuity, not are they (worse yet) mere arbitrarily designated segments of that continuum. Therefore, it is impossible to explain the history of life purely in terms of an adaptive continuum of diversity and to account for the existence of species at the same time. Put the other way around, considering species as real units in nature prevents the reduction of the history of life to the principles of adaptive change in the phenotypes of organisms—where adaptation is construed as resource competition rather than reproductive competition. In other words, there is a useful distinction to be drawn between organismic adaptations for reproduction and adaptations for matter–energy transfer pertaining to somatic maintenance and growth. It is significant, I believe, that Dobzhansky wrote that passage in 1951; nothing quite so bald appeared in his earlier two editions.

## PUNCTUATED EQUILIBRIA

"Punctuated equilibria" (Eldredge 1971; Eldredge and Gould 1972; Gould and Eldredge 1977—see Eldredge 1985 for a general account), coming along at about the same time as Ghiselin's (1974a) and Hull's (1976) more general ontological arguments, adds empirical support to the notion that species are spatiotemporally bounded entities.

The concept of punctuated equilibria is based on the observation that in terms of those phenotypic characteristics that change when an ancestral species gives rise to a descendant (the "species-specific *differentia*") the change occurs *relatively* rapidly when compared with the total longevity of both ancestral and descendant species. Typical values, for marine invertebrates, are species durations of from five to as many as ten or more million years, while the time required for anatomical change to occur is on the order of five to fifty thousand years (Eldredge and Gould, 1972; Gould and Eldredge 1977; Stanley 1979). The empirical basis for punctuated equilibria is the observation that morphological change appears to be episodic to a significant degree. Because of the extreme stability of such morphological packages, and because they frequently can be shown to overlap in time, they are interpreted as biological species, allopatric speciation is invoked to explain the origin of descendant from ancestor, and the further conclusion is drawn that (given the validity of these two assumptions) anatomical change seems in general to be concentrated in speciation "events."

Eldredge and Gould (1972, p. 111) identify a paradox of sorts: lack of concerted, directional morphological change within species was held to conflict with patently among-species directional evolution (i.e., trends). This argument led directly to the present thicket of "species selection" models—to which I return shortly. The implications of punctuated equilibria on the ontological status of species are what concern me here.

It is widely alleged that species, when defined as reproductive communities, cannot be recognized with rigor in the fossil record (see, e.g., Levinton and Simon 1980 for a typical argument). Eldredge and Cracraft (1980, chapter 3) examine the problem of species recognition in detail. Absence of crucial soft-tissue anatomical, physiological, and behavioral information results in lumping too many species together. Vrba (1980) points out that Paterson's (e.g., 1978 and later references) focus on species mate recognition systems (SMRS) has definite implications for species recognition in the fossil record. The eastern North American leopard frog, the one we all learned as *Rana pipiens*, is now known to be several species, distinguished primarily by their different vocalizations. Yet all passed as a single species for years, and obviously would be so regarded forever if they were known solely from fossils. Following Vrba's (1980) lead, we can be most confident that paleontological species conform to actual species when parts of Paterson's SMRS are directly preserved in fossils, as in the horn morphology of African antelopes (Vrba 1980, 1984a).

And paleontologists share the difficulties in dealing with allopatric (and allochronic!) populations of a polytypic species so clearly spelled out by Mayr (1942, pp. 151 ff.; see my earlier discussion in chapter 3). There is simply a danger of oversplitting.

Yet the fossil record of metazoans, at least, is essentially a complex sequence of stable anatomical packages. A Paleozoic bedding plane may reveal more than twenty different "kinds" of invertebrate shells, as easily sorted as the mollusks on a modern beach. The point is that those discrete packages in nearly all cases have *non*trivial temporal ranges. In the rocks I personally know best—the "Hamilton group" of eastern North America (see Eldredge 1985 for extended discussion)—there are perhaps as many as two hundred species of invertebrates, and nearly all persist for the six- to eight-million-year interval those rocks represent.

The point is not to argue for the prevalence of rigid, straightjacketed stasis. Stasis itself raises interesting problems (best addressed, I think, in terms of hierarchy). But there *are* these (relatively) stable, conservative anatomical entities that seem to have beginnings, histories, and terminations. They are the very stuff of the fossil record. They are precisely what one would expect to see if species were construed as spatiotemporally bounded entities. Thus, the empirical basis of punctuated equilibria agrees very well with the simple idea that species are individuals. If, how-

ever, these conspicuous and discrete anatomical packages are not species—if, indeed, it is improper to speak of such collections of similar
organisms as even constituting packages of any sort—then, of course, the
implications of punctuated equilibria for an ontology of species are not
demonstrated. If, though, there is a case to be made for species as spatially
bounded entities at any one point in time (and remember that Mayr
explicitly, if only partially, made precisely that claim in 1942), the empirics of the fossil record strongly suggest that they are temporally bounded
as well. I have discussed the connection between species, speciation, and
the fossil record at greater length, albeit in more general terms, in my
*Time Frames* (1985).

There are other features of species (and, indeed, of all classes of biological entities that qualify as individuals in particular instances) that stamp
them as individuals. All individuals, for example, need some "glue"
imparting internal cohesion (see chapter 6 herein). But explicit invocation
of such a revamped ontology to support a claim of hierarchy is to be found
so far only in Eldredge (1982b), Eldredge and Salthe (1984), and Vrba
and Eldredge (1984). It forms the basis of the remaining chapters of this
book. The main focus of hierarchical approaches to evolution in recent
years resides in the by now familiar flap over levels of selection, into
which the ontological status of biological entities has entered—but not, I
think, entirely correctly or even from the most interesting point of view.

## LEVELS OF SELECTION: PROMULGATIONS OF HIERARCHY

Some of the recent spate of group selection literature is decidedly pro.
Wade (1978) and others have all concluded that there is something to
*some* of the versions of the notion, yet it is clear that many would agree
with Dawkins when he says (1982, p. 115) that "even the staunchest
group selectionist would agree that the individual organism is a far more
coherent and important 'unit of selection.'" Group selection is still seen
as a secondary phenomenon for the most part, and Williams's contention
(see chapter 4 herein) that it is essentially an onerous concept and only
necessary when shown to conflict with an argument based on pure organismic selection remains the consensus.

Yet higher-level selection continues to attract attention. Vrba and Eldredge (1984) have reviewed a number of somewhat conflicting versions
of "species selection"—a term introduced in its latest form by S. M. Stanley (1975a; also 1979). Implicitly treating species as individuals (or as
"particles" in the interestingly parallel usage of Raup, e.g., 1977), Eldredge and Gould (1972) pointed to the aforementioned paradox. There is
a theory—directional artificial selection—that accounts for linear change
within populations of experimental organisms. The inference is that linear
natural selection effects changes in the adaptations of organisms. Thus, we
have another theory readily available—prolonged directional natural

selection ("orthoselection"; see Simpson 1944)—to explain directional changes that paleontologists observe through geological time. The paradox arises from the empirics of the situation. We observe virtual stasis *within* the packages we interpret as species; change *among* species is accomplished in relatively brief spurts. Suppose Wright (1967; also Mayr 1963) is correct in seeing an analogy between mutation on the one hand and speciation on the other; morphologic change at speciation is random with respect to the direction of the trend.[4] We observe a progression of change, and the change is even, perhaps, to be construed as adaptive. Yet natural selection is not constantly working within a lineage to effect the long-term change—the easiest, most direct application of selection to explain trends, and the model utilized by the synthesis. Yet if natural selection *is* effecting among-species change at speciation, the change is not always in the direction of the trend. Thus, it is concluded, there must be some sort of higher-level culling process going on. There must be an ordering mechanism that selectively preserves or favors those species that are in fact moving in the direction that we characterize as a trend. It is this phenomenon that Stanley (1975a) termed species selection.

It is to be noted at once (and with the benefit of more than a decade of hindsight) that a pattern, a general class of similar historical events, is alleged to exist: trends in anatomical change occasioned by the differential origin or persistence, or both, of anatomical packages interpreted as species. Trends do seem to exist as phenomena (see chapter 7 herein), though there is some skepticism about how close to reality was the "thought experiment" that saw anatomical change between species as wholly random with respect to long-term trends. (For example, in hominid evolution a case can be made for an interspecific pattern of accumulation of larger-brained hominids through time, yet I am unaware of any apparent reduction—or even simple retention—of absolute brain size at any alleged speciation event in hominids over the past four million years.)

But the mere existence of such a pattern, though it does indeed *falsify* the simplistic extrapolation of natural selection (in the form of orthoselection) as an adequate explanation for trends, does not thereby establish that some alternative or additional higher-level selective process is at work. Indeed, the main value of all the recent discussions of species selection seems to be that patterns of differential origin or survival of species are finally being taken as suitable for serious theoretical analysis. In other words, the main value is the (largely) implicit status of species as *individuals* underlying all the discussion.

Vrba (1984b; also Vrba and Eldredge 1984) has pointed out the manifest confusion permeating most discussions of species selection. I follow her analysis here. There seem to be three major sources of muddling in the debate as it has unfolded. The first is simply the confusion of the pattern with the process, the assumption that just because there is indeed a pattern of differential species origin and persistence, that it bespeaks ipso facto a group-level process of selection. The second source has been the definite tendency in the earlier literature (e.g., Eldredge and Gould 1972) to identify the notion of species selection predominantly (if not solely) with differential species *survival*—(e.g., species' "deaths"). Such a concept is similar, of course, to the older version of natural selection, where the death of an organism, and not merely its genetic death through failure to reproduce, was the prevailing metaphor. The modern version of natural selection, of course, stresses differential reproductive success (itself a somewhat one-sided view; Eldredge and Salthe 1984 resurrected the actual differential births *and deaths* of organisms as the more general pattern that natural selection underlies—something of a return to an older view). The point is that patterns of differential species births and deaths belong to the rubric of species selection. But, as Vrba (1980, 1983, 1984b) has been arguing, such patterns may arise from lower-level causes. We simply may not need a species-level selective mechanism to account for the patterns of differential species births and deaths that we do indeed see empirically.

That brings us to the third source of confusion. Some theorists (e.g., Arnold and Fristrup 1982, though they are inconsistent on this very point) have clearly recognized that species selection must involve what Williams (1966) terms biotic adaptations. Most others, however, have overlooked this rather obvious point (myself included, e.g., Eldredge and Cracraft 1980). The only kind of trend we can identify with any degree of confidence in the fossil record is, of course, morphological—and here we are patently dealing strictly and solely with the anatomical features of *organisms*. This is not to deny the possibility of species-level adaptations (and "emergent properties"—see below) but merely to acknowledge the obvious. The pattern of differential production and culling of metazoan species that empirically underlies most (I believe all) trends in the fossil record always involves simple phenotypic characteristics of organisms.

Vrba (1984b and personal communication) insists that consistency demands that species selection treat of "species characteristics," emergent properties that are not the mere sum-of-the-parts attributes of the organisms belonging to the species. I must agree. We can explain the differential survival (as an example) of species within a lineage with recourse to nothing but natural selection at the pure organism level (i.e., acting on the phenotypic properties among organisms within populations, so long as those phenotypic properties are under genetic control). Indeed, Vrba's (1980, 1983) "effect hypothesis" is designed to achieve precisely this sort of explanation. Put another way, imagine two species of hyrax living (as they do) on the kopjes of eastern and southern Africa. The bush hyrax climbs up into bushes and trees, eating leaves, while the rock dassie stays below and munches grass. Bush hyraxes are known to be able to eat grass as well. Suppose an African drought continues indefinitely; depending upon the exact nature of the drought's impact on the vegetation, there will be substantial change in the African biome. We might expect that one—or both—of these species will become extinct. If both, one will probably precede the other. Suppose one becomes extinct and the other survives until lusher times are restored. Is this species selection? Certainly not; all we need to know are the food requirements of the organisms within each species in order to understand why one species survived and the other did not (or indeed to predict what will happen given various scenarios of community change). This is but one (extremely simple-minded) way in which the effect hypothesis can work. The point is that, from a paleontological perspective, for the most part we can deal only with organismic attributes.

There remains, of course, the possibility that there truly are species-level "emergent" properties, which can be construed as species-level adaptations under the control of species selection. Sexual reproduction, for instance, generally requires more than a single organism. It may be argued that sex is a group-level character. Yet I think that Ghiselin (1974a, p. 540) is right: the forma sexualis, as he puts it, remains a class within a species. The differences in sexual reproduction between species—differences in "life history strategy" (the entire $r$ and $K$ syndrome; see Stearns 1976 for a useful review)—materially affect the potential for survival of a species and its potential for producing more species. Yet such differences boil down to the anatomical, behavioral, and physiological properties of the two classes of organisms within each of the species being compared: the males and the females.

Dobzhansky (1937a) saw allelic frequencies as a group-level property. The point is important. Relative frequencies are simply the additive properties of the members of the group. The food preferences of the hyraxes can be construed in terms of such group-level frequencies. Thus, the differential success of entire groups dependent upon the frequency of this or that organismic adaptation can indeed be construed as group-level. But we are still dealing with the properties of organisms themselves, and in general frequencies of the properties of lower-level individuals which are parts of a higher-level individual simply do not make convincing higher-level adaptations.

The same may be said, for instance, for the niche-exploitation parameters of eurytopy and stenotopy, currently a favorite concept in many discussions of differential species origin and survival (e.g., Eldredge 1979; Vrba 1980; Eldredge and Cracraft 1980). And we ask in what sense a species can be said to be eurytopic as a species-level attribute. There are two ways for a species to be eurytopic (Fox and Morrow 1981). First, all of its component organisms may be physiologically eurytopic, in which case we are dealing simply with an organism-level feature that happens to hold for all members of a species. Alternatively, species-level eurytopy may involve some form of polymorphism, where organisms of each class focus on some part of a physiological spectrum, and the entire spectrum of such physiologies is the simple accumulation of the several classes. (Fox and Morrow think that the latter is a more common phenomenon than is usually acknowledged.) The latter situation is exactly like Dobzhansky's argument about gene frequency; such additive, spectrum-based eurytopy is still not the sort of species-level property we truly need in order to speak of species-level adaptation as clearly distinct from organism-level adaptation.

Indeed, I can think of no utterly convincing species-level adaptations. This is because I can think of no utterly species-level properties except for (1) their boundaries and (2) their source of internal cohesion. The former *are* species properties (as Fowler and MacMahon 1982, p. 482, seem to agree); temporal and geographic boundaries of species are *not* intrinsic features of organisms. But neither is there a mechanism for some form of hereditary transmission for such properties, whether involving organisms or some other as yet unknown and unsuspected form of hereditary transmission. And the source of internal cohesion for a species is simply the ongoing reproductive plexus of its component organisms. Neither class of such species-level properties can be construed as species-level adaptations.

There is a risk here of being too formal, too restrictive. The synthesis simply does not address the comings and goings, the differential origins and disappearances, the differential success of species. Yet this seems to be very much a dominant theme in the history of life. Thus, the entire possibility of species selection could be safely ignored in the synthesis. This remains the case even though I feel that for the sake of consistency we shouldn't be calling it species selection if what we are discussing really does not entail species-level adaptations in any real sense. And there is another at least equally severe objection to the entire notion of species selection that I develop in the next chapter.

Again, it is not the argument that species selection (or any other level of selection other than natural selection itself) must exist that leads us to hierarchy. It is the ontology of the thing. The synthesis treats species as temporally at least potentially unbounded, hence classlike entities. Taxa of higher rank are treated even more sloppily. Such treatment allows, indeed *is necessary for*, the bald extrapolation of natural selection within unbroken lineages to explain the supposed patterns of phenotypic (organismic) diversity through time. Such a theory ignores the real packaging of those phenotypes (and underlying genotypes) in nature. But when all is said and done, the patterns we address still involve phenotypic diversity. Therefore, we require a theory that recognizes the nature of the packaging of information—as Hull (1980) says, not just genes and organisms, but demes, species, and (I would add) monophyletic taxa. But the theory also must continue to address the problem of phenotypic diversity. As Williams (1966) says, it is not impossible that there are other sorts of adaptations involving other sorts (i.e., levels) of selection, but it may be that these are not really important. More interestingly, such adaptations are really almost beside the point because they do not focus on the problem already at hand: the diversity and discontinuity of life. It is the discontinuity, the packaging of that diversity in a manner rather more complex than the synthesis has acknowledged, that provides the key to a more accurate depiction of the evolutionary process. Packaging of information—the information that lies at base in the genome—is the basis of recognizing one of the two process hierarchies in evolution.

## OF REPLICATORS AND VEHICLES: SELECTION AND PACKAGING

Dawkins (1976, but especially 1982) and Hull (1980) have treated levels of selection in rather similar ways and in so doing have made explicit statements about the ontology of various sorts of biological entities. Hull (1980) distinguishes between two classes of entities: *replicators* (he cites Dawkins 1976 for the term) and *interactors*. A replicator is "an entity that passes on its structure directly in replication." An interactor is "an entity that directly interacts as a cohesive whole with its environment in such a way that replication is differential." Thus, selection is "a process in which the differential extinction and proliferation of interactors cause the differential perpetuation of the replicators that produced them" (Hull 1980, p. 318). Both replicators and interactors were chosen as neutral descriptors: Hull's aim was to evaluate the possibility of multilevel selection. Though he comes to no definitive conclusion, the prospects for higher-level selection (i.e., higher than natural selection) are not good; genes clearly replicate (indeed, as Dawkins suggests, they are the only entities that do). Genes make organisms and clearly organisms interact with their environments, yielding, in Hull's vision, natural selection. Going up the scale of replicative entities, Hull concludes (1980, p. 324): "In sum, replication seems concentrated at the lower levels of the organizational hierarchy, occurring usually at the level of the genetic material, sometimes at the level of organisms and possibly colonies, but rarely higher." As for interactors, Hull concludes (1980, p. 327): "In sum, entities function as interactors at higher levels of organization than those at which replication occurs, at least at the level of colonies, possibly at the level of populations, but probably at no higher levels."

Focus on selection has led analysts, particularly Hull and Dawkins, ineluctably to consider replicators. Dawkins sees only genes as replicators (strict, faithful self-copiers) and all other units (organisms, demes, species) as mere *vehicles* for the replicators. He writes:

A vehicle is any unit, discrete enough to seem worth naming, which houses a collection of replicators and which works as a unit for the preservation and propagation of these replicators. I repeat, a vehicle is not a replicator. A replicator's success is measured by its capacity to survive in the form of copies. A vehicle's success is measured by its capacity to propagate the replicators that ride inside it. The obvious and archetypal vehicle is the individual organism, but this may not be the only level in the hierarchy of life at which the title is applicable. We can examine as candidate vehicles chromosomes and cells below the organism level, groups and communities above it. At any level, if a vehicle is destroyed, all the vehicles inside it will be destroyed. Natural selection will therefore, at least to some extent, favour replicators that cause their vehicles to resist being destroyed. In principle this could apply to groups of organisms as well as to single organisms, for if a group is destroyed all the genes inside it are destroyed too.

Vehicle *survival* is only part of the story, however. Replicators that work for the "reproduction" of vehicles at various levels might tend to do better than rival replicators that work merely for vehicle survival. Reproduction at the organism level is familiar enough to need no further discussion. Reproduction at the group level is more problematic. In principle a group may be said to have reproduced if it sends off a "propagule," say a band of young organisms who go out and found a new group. The idea of a nested hierarchy of levels at which selection might take place—vehicle selection in my terms—is emphasized in Wilson's (1975) chapter on group selection (e.g., his figure 5.1). (Dawkins 1982, p. 114).

It is difficult to disagree with much of what Dawkins says here. Surely it is true that as one ascends an organizational hierarchy of entities (his vehicles) anything resembling replication becomes progressively more remote. The key notion that I shall develop in the next chapter is that all of these entities do indeed reproduce. They do not necessarily make more of themselves; *they make additional entities of like kind*, the sine qua non criterion for inclusion in the genealogical hierarchy. Ascending the scale from genes through species, resemblance between parent and offspring becomes progressively less faithful. Alexander and Borgia have made a similar point:

Units or groups such as species, then, may be established through individual or genic selection, yet persist or fail as a result of competition with other species—hence, through a kind of group selection. Dawkins (1976) has denied that differential species extinction can properly be termed group selection because, as he puts it, species are not "replicators." But of course they are: Species give rise to species; species multiply. (Alexander and Borgia 1978, p. 456)

I would only question their conclusion that the multiplication of species amounts to replication, preferring myself the more neutral descriptor *reproduction*.

But I fear that emphasis on selection and on replicators (even on interactors, Hull's useful addition to the discussion) leads us to throw the baby out with the bathwater. I find the vehicle concept far more compelling than Dawkins evidently does. For if it is true that replication fidelity decreases up the scale of entities (to the point where the term only really describes the reproductive activities of genes), the inverse is true for another vital aspect of these entities: their *permanence*.

Williams (1966) and Dawkins (1976, 1982), for all their talk of the potential immortality of genes, have always quietly acknowledged that it is the *information*, not the genes themselves, that is potentially immortal. I have already (p. 107) cited Dawkins on the progressive ephemerality of higher-level units. Species, for example, hardly seem to exist in any meaningful sense—to Dawkins, that is. But with this newer ontological position, where we abandon this "commonsense" view (to use Hull's term), we see that precisely the opposite is obviously true: genes are shorter lived than organisms, organisms shorter lived than demes, and species shorter lived than monophyletic taxa of higher rank. Life, after all, is the least ephemeral of all these entities. And there is a genetic resemblance of a basic and major kind that unites all life forms.

I am simply suggesting that there is much from a "selfish gene" point of view to exploring these nested vehicles, these entities that form the genealogical hierarchy. (Hull's interactors belong to the parallel ecological hierarchy, also discussed in detail in the next chapter.) Replication falls off as longevity increases in these entities. And longevity of these units would seem, on the face of it, to be heavily involved in the survival of those selfish genes—not from the standpoint of the individuals themselves, each particular codon, but from the standpoint of the information that each contains. It is the survival of that information (instructions for products; information about the environment in the broadest sense) that the evolutionary game seems to be all about, at least from a genealogical perspective. Species might simply be stable entities, packages of information, maintained albeit in more dilute form (but for longer periods of time) than in exact replicants. In short, pursuit of multilevel selection has indeed gotten us thinking in hierarchical terms. But it is a blind alley; the very implausibility of group (meaning, really, "individual higher than organism") selection has blinded us to the true nature and significance of hierarchy.

Dawkins writes (in the passage just quoted) of the "survival of the vehicle." I will return to this complex, knotty problem after a formal presentation of the structure of the two process hierarchies and some aspects of the content of a hierarchically organized evolutionary theory. We need to know, ultimately, *why* the information ensconced in the genotype is packaged in all those entities. Dawkins asks (1982) why do we have organisms? But why do we have demes? And, as Dobzhansky asked, why do we have species? And what are the consequences of this organization for the patterns of maintenance and change of this information (i.e., evolution) and the conundra that Hull lists (1980, p. 311): "the presence of so much genetic heterogeneity in natural populations, the prevalence of sexual forms of reproduction in the face of an apparent 50% cost of meiosis, and the difficulty of explaining how selection can operate at higher levels of organization"? An explicitly formulated, hierarchically structured evolutionary theory will give us the wherewithal to integrate ecological, molecular, developmental, and macroevolutionary theories more effectively with the core neo-Darwinian paradigm—more effectively, that is, than the synthesis in its past and present manifestations has been able to achieve. Hierarchical theory will also give us the conceptual tools needed to tackle the sorts of residual problems Hull has identified. The reason is simple: the biotic world is composed of a wider variety of individuals than explicitly addressed by conventional evolutionary theory, the synthesis and its subsequent tradition. The synthesis considers genes, organisms, demes, and (partially) species. This is a critical and obvious portion of the genealogical hierarchy. We must now add the remaining elements of that hierarchy and recognize as well the existence of the parallel ecological hierarchy. It is the interactions of the nested individuals within each hierarchy that has given us the history of life on earth—evolution.

## NOTES

1. I agree with Gould and Vrba (1982) that, strictly speaking, we should be using the term *aptations* to refer to organismic attributes that perform a specific function, whether shaped by natural selection for that particular function or not. *Adaptations* are those aptations specifically fashioned by natural selection to serve the use that we observe. This is the sense in which I use the term adaptation throughout the remainder of this book.

2. Dawkins (1982, p. 108) also cites some of the "punctuational" literature (e.g., Eldredge and Cracraft 1980, p. 269) as proof that we do not, after all, abandon the notion of adaptation through natural selection—implying that there was confusion on that point in the first place.

3. I freely confess that after reading Ghiselin's paper when it was first published, I could not see its relevance to evolutionary theory, and I thought the idea, to the extent that I grasped it, was trivial, hardly warranting so dramatic a description as radical. I was wrong. Ghiselin's book (1974b) was an early attempt to follow out the implications of this revised ontology—and to see evolution as a dual concern of economics and sex. This is certainly similar in spirit if not so much in particular content to the proposition (Eldredge and Salthe 1984; chapters 6 and 7 herein) that it is the interactive relationship between the genealogical and ecological hierarchies that is the evolutionary process.

4. Of course, morphological change at speciation is heavily constrained by the preexisting phenotypes and genotypes of the organisms sampled from the ancestral species that are the founders of the new species. All that is postulated here is an independence of that change from the single direction of change that is retrospectively seen as constituting a linear trend in the entire lineage of which each species is but a single part.

# 6

# The Evolutionary Hierarchies

"Theory generally should not be an attempt to say how the world is. Rather, it is an attempt to construct the logical relations that arise from various assumptions about the world." (Lewontin 1980, p. 65)

Real-world hierarchies are usually treated as patently evident manifestations of nature, a fact that has somehow rendered them trivial. Familiarity breeds contempt. Everyone knows that atoms are the building blocks of molecules, small molecules combine to form the huge molecules of living systems, organelles and other structures composed of such molecules make up cells, cells link up to form tissues, tissues do likewise to form organs, thence organ systems, and finally we arrive at the integrated soma of the organism. And organisms are associated to make populations, or demes, or species. The very names of the subdisciplines of biology (especially as conceived fifty years ago) reflect this organization. Though molecular biology is a recent arrival, physiological genetics, cytology (and cytogenetics), histology, and physiology nicely recognized the components of the somatic or organismic hierarchy.

Thus, the *somatic* hierarchy retains a heuristic value still reflected in general biology texts. The molecules-to-organism hierarchy offers a handy way of organizing biological knowledge, information about living systems. That nature itself is organized in such a fashion seems to have slipped to secondary significance. It is, though, common to view the sequences of codons that compose functional genes, as well as other organizational features of DNA molecules, as constituting the retention of information. One major biological hierarchy, the *genealogical* hierarchy, is evidently a hierarchy of information. It is as if the pragmatic, heuristic, epistemological aspects of biological hierarchies, providing us with a handy way of organizing, summarizing, and communicating what we think we know about biological systems, serve to obscure the significance of hierarchical organization to the very workings of biological nature.

All of this does not deny that there is an extensive literature on hierarchies, a multidisciplinary literature that includes a long, if episodic, history within the realm of biology. The analysis I develop here surely does not arise from a vacuum. Yet the current resurgence of hierarchical outlook on evolution reflects more, I think, a return to an alternative way of looking at nature, a way dictated by a pattern of organization of nature that is there for all to see, than it does the thickening of a continuous intellectual strand that connects us with earlier interests in hierarchy both within biology and without. Dobzhansky's (1937a, p. 11) three levels, of course, continued to be recognized, but he, along with everyone else, dropped the explicit development of a hierarchical context in treatments of evolutionary processes. Current interest in hierarchies from an evolutionary point of view rests on the explicit ontological reformulation of the status of various sorts of entities taking part in the evolutionary process, as discussed in part in the preceding chapter. I say this despite my acknowledgment that the explicit rationale conventionally given for considering a hierarchically structured evolutionary theory is a claim about *process*, namely that selection takes place on more than one level. If the discussion in chapter 5 establishes anything, though, it is that such treatments of process are intrinsically bound up in the deeper issues of the ontological status of the entities said to be selected. Given the dubious status of most multilevel selection models, moreover, the stronger claim that evolutionary theory needs to be formulated in hierarchical terms actually rests on the notion—the observation, really—that the entities that take part in the evolutionary process are themselves hierarchically organized. We will then be in a position to recognize processes (e.g., births and deaths and factors biasing those births and deaths) at the various levels. Moreover, above the organism level biological entities appear to be segregated into two functional classes: those individuals that take part in the economic side of nature (i.e., matter–energy transfer) and those that are engaged instead in the conservation and transmission of (hereditary) information.

## WHAT ARE HIERARCHIES?

Hierarchies usually convey the notion of rank—a series of levels, such as the variously ranked grades of authority in the armed services. Generals outrank majors. Each rank is a class; the persons assigned to each rank are obviously individuals, but neither the classes nor their individual components include the classes or individuals of the lower ranks. In contrast, hierarchies that seem to me relevant to the evolutionary process are inclusive: organisms are composed of organs, which are composed of tissues, which are composed of cells, and so on. Four of the five hierarchies I discuss here—the genealogical, ecological, taxic, and somatic hierarchies—consist of a series of individuals $(I_n)$ such that each $I_n$ is composed minimally of one but actually of many individuals of the next lower order of magnitude $(I_{n-1})$. The only exception occurs at the first level (i.e., lowest) individual, which is not recognized to consist of other, even still lower-level individuals (even though it very well may be decomposable into such smaller individuals). The fifth hierarchy—that of homology—is only partly inclusive, as I shall discuss in chapter 7.

It is common to speak of such hierarchies as nested (e.g., Eldredge 1982b). It is incorrect, however, to call them nested sets, as I did in that paper and have done elsewhere. "Sets" obviously are classes. The trick in seeing these biological hierarchies correctly is to view them as nested but not to think of the higher levels merely as aggregates of lower-level individuals—that is, as sets. The higher-level units are themselves individuals, though not ipso facto, as the ontological status of each putative individual needs to be independently established.[1] Thus, for each hierarchy broached in the following pages I present a formal argument for the view that at all the levels (classes) of entities included in the hierarchy, the entities that occupy those levels are, in each particular instance, to be construed as individuals.

Inclusive hierarchies of the sort I consider here have several other important general properties as well. The nesting of discrete individuals into larger individuals sets up a system of levels. We might expect that dynamics within each level will themselves be discrete, not directly connected to dynamics of change within the adjacent lower and upper levels. For example, Stanley (1975a) described the process of species sorting that he termed species selection with the useful term *decoupling*: species sorting goes on among species within a monophyletic clade, independently from the lower-level process of natural selection (or interdemic selection).

Yet, from another point of view, what goes on at one level within a hierarchic array will affect processes and events at other levels. For example, variation is introduced by mutation from Dobzhansky's (1937a) lowest level and injected into the higher level of the population, where drift and selection take over as dynamics. Vrba and Eldredge (1984) generalize this phenomenon of the introduction of variation from lower levels within the entire genealogical hierarchy. Eldredge and Salthe (1984) speak of constraints where a lower level affects processes and events at the next upper level. They speak of boundary conditions when the causative flow is reversed, and conditions of the next upper level determine to some degree what goes on in the next lower level. The interactive nature of the various levels in such an inclusive hierarchy may be many and varied. There is both upward and downward causation (Campbell 1974; Vrba and Eldredge 1984) which may transcend far beyond the interactive effects between the individuals of one level and their nearest (higher and lower) neighbors—the basic triad of Salthe (1985) and Eldredge and Salthe (1984). Thus, in general we are looking at systems that have a full measure of independence of existence of levels based on the internesting of discrete individuals. There is as well a degree of autonomy of event and process within each level, and yet we would also expect some interactions, or interactive effects, among levels.

It is possible to be far more precise and specific about the general nature and structure of all hierarchies. Salthe (1985) has provided just such a detailed discussion (see also Eldredge and Salthe 1984). My approach here, instead, will be to develop the several hierarchical systems as cogently as possible. It is, I feel, for the most part a task of description. I shall then analyze the nature of the relationships among levels, and then between hierarchies, without recourse to a formal, theoretical set of expectations beyond the few simple precepts outlined above.

Much has been said about hierarchies in an evolutionary context in the past half-dozen years or so, yet relatively little concrete has been done about them. Certainly the species selection literature not only foreshadows hierarchy but actually has begun calling explicitly for hierarchies (cf. Arnold and Fristrup 1982; Gould and Eldredge 1977; Eldredge and Cracraft 1980; Gould 1980a, 1982b). In an important paper, Alexander and Borgia (1978) review group selection models within an explicitly hierarchically structured ontology—the "levels of organization of life." Hierarchy theory has become particularly prominent in the literature of ecology (cf. Valentine 1973; Allen and Starr 1982; MacMahon et al. 1978) and in the literature of developmental biology as well. I cite these and other works in relevant sections below; therefore, I will forego a separate historical review of the recent literature. In any case, Salthe (1985) provides a comprehensive survey of the literature. The particular scheme adopted, defended, and analyzed in the following pages has its direct ancestry in three specific discussions: Eldredge 1982b, Vrba and Eldredge 1984, and Eldredge and Salthe 1984—and, of course, references cited therein.

## SOME BIOLOGICAL INDIVIDUALS

What evolves in evolution? We return to the list of candidates, biological entities, whether construed as individuals or classes or something of a hybrid (cf. the usual conceptualization of species within the synthesis): base pairs, codons, functional genes, pseudogenes, families of repeated noncoding DNA, chromosomes, cells (with organelles, some with their own DNA, e.g., mitochondria and chloroplasts), organisms (with their own internal hierarchical arrangement of entities), demes, populations, communities, regional biotas, species, and monophyletic taxa. The list is hardly exhaustive; there is a variety of other classes of elements already known that comprise the genome. Indeed, there seems to be a hierarchy of control intrinsic to the genome itself. Also missing from the list are grades (see chapter 4 herein), which are morphological abstractions (classes) for the most part, becoming less classlike only in a phylogenetic sense; when anatomical grades coincide with monophyletic taxa, grades are coincidentally individuals. Grades are patterns, gross-level generalizations of the results of the evolutionary process which are indeed properly addressed by evolutionary theories. But grades themselves, purely construed, do not play an active role in the evolutionary process because they do not really exist as individual entities.

The list is thus minimal rather than exhaustive; there may well be additional classes of entities that do belong to it and that I have omitted out of simple ignorance or the misapprehension that entities such as grades cannot be construed as individuals. Demes constitute a counterexample. As I have already remarked, Dobzhansky (1937a) relied heavily on the notion of demes (Wright's "colonies"), but subsequently, until recently, evolutionary theory in general has gotten along perfectly well without paying much more than lip service to the very existence of demes, hence without truly incorporating Wright's "shifting balance theory" fully into the paradigm (see Salthe 1985). If one doubts the existence of demes, one may omit them from the list of entities that we associate with evolution. In a *structural* sense, it does not matter; we can add and subtract entities from the hierarchies—the two nested hierarchies that fall out of this list—and still have a hierarchical structure. What changes, of course, is the specific *content* of the process theory that we tack onto that structural framework—no small matter indeed. From the standpoint of the content of evolutionary theory, of course, we had better get our ontology straight and worry about such questions as the actual existence of demes in nature. Putting the issue in a slightly different light, consider the very real possibility that demes exist within some species but not within others. The specific theory of within- and among-species evolution will accordingly differ, but the conceptual framework remains hierarchical: demes are removed from the hierarchy when they are not a true description of nature, and we go from organisms directly to species in the hierarchical system.

## Some Basic Sorting

Simple inspection of the list of biological entities that we tend to associate with evolution quickly divides them into two classes: those entities that are patently ecological and those that seem more related to ancestry and descent, to genealogy. We immediately get two lists: regional biotas, communities, populations, organisms, cells, and proteins on the one hand; and monophyletic taxa, species, demes, organisms, chromosomes, and the elements of the genome (genes for short) on the other hand. Note that organisms are included in both lists, a circumstance that is at once somewhat problematical and yet perhaps the crucial link between the two hierarchies (see chapter 7 herein).

And the lists are the class levels of two hierarchies, inclusive ones at that. It is the structure and content of both the ecological and the genealogical hierarchies which, in my opinion, hold all of the ingredients of the evolutionary process. There are processes within each (for example, replication of the genome and the dynamics of predation) which, when taken alone, constitute some small portion of the evolutionary process. And there are interactive effects among levels within each hierarchy, as when genomic constraints limit the number of different forms that allelic mutations can take, thus limiting the range of variation that can be expressed within a population on which natural selection can act. But it is the cross-interactions between individuals of the two hierarchies that integrate all these entities and processes into evolution. The interplay between the two process hierarchies—genealogy and ecology—produces the evolutionary patterns we are all familiar with, patterns that include the taxic hierarchy, the complex hierarchy of homology, and the parallel linear ecological and genealogical histories of life that span the past three and a half billion years. I pursue the nature of such interactions in the following chapter. The remainder of the discussion here is devoted to a detailed explication of the genealogical and ecological hierarchies.

## THE GENEALOGICAL HIERARCHY

What makes the elements, the entities, of the genealogical hierarchy— the genes, chromosomes, organisms, demes, species, and monophyletic taxa—*individuals*? As we have seen (chapter 5), individuals are spatiotemporally bounded, thus historical entities. They have beginnings, histories, and terminations. Their boundedness, one might suppose, devolves from some combination of material properties that actually form boundaries (such as the ectoderm of vertebrate organisms or the double-walled membrane that defines the outer limits of the nucleus of the eukaryotic cell). But such boundedness typically arises as well from a sort of constituent "glue" that associates the subparts of an entity, giving it cohesion, unity and integrity characteristic of individuals. It is on this putative glue that I shall base my case for the ontological status of most of these various sorts of entities as individuals. Moreover, there is a class of such glue common to all members of the genealogical hierarchy; this class, in fact, becomes a necessary (but not wholly sufficient) criterion for membership in that hierarchy.

That "glue" is simply "more-making." Every element in the genealogical hierarchy (save one) is capable of making more individuals of like kind: codons make more codons, genes in general make more genes, chromosomes make more chromosomes, cells make more cells, organisms make more organisms, demes make more demes, and species make more species. "Individuals of like kind," of course, constitute a class, so it is appropriate at this juncture to acknowledge that the list of genes, chromosomes, organisms, demes, species, and monophyletic taxa is just that: a linear list of classes (categories). Particular instances of these classes are the individuals of the evolutionary process—the gene in one particular cell that codes for alcohol dehydrogenase, the species *Homo sapiens*.

What makes these entities individuals, however, is not their own capacity to make more entities of the same class but the fact that the next-lower-level constituents are also making more of themselves. Take species, for example. I will pursue the individuality of species in greater detail shortly, but the outline of the argument is simple and clearly illustrates the general point. The definition of *species* that I follow here (see chapter 5) is a variant version of the "biological species concept" developed especially by

Dobzhansky (1937a,) and Mayr (1942). Species are seen as reproductive communities, "within which there is a parental pattern of ancestry and descent, beyond which there is not," as a part of a generalization of the biological species concept would have it (Eldredge and Cracraft 1980, p. 92). Leaving demes aside for the moment, the simple reproductive activities of sexual organisms, activities that continue to produce more sexual organisms, are what keep a species going. Both spatial boundedness and temporal continuity, at least in this case, of a higher-level individual—a species, in this instance—derive from the simple process of more-making by the next-lower-level constituent individuals—in this case, sexually reproducing organisms.

The genealogical hierarchy itself is not infinite. It is bounded both on top and on bottom ("truncated" in the language of Salthe, 1985). Base pairs appear to constitute the lowest level of molecular organization engaged in the process of making more of themselves. Replication, the sine qua non of life, is after all a special feature of DNA not generally found in other molecular systems. The lower boundary of the genealogical hierarchy, then, is set by the smallest elements that we might call individuals that are implicated in the regular processes of reproduction and the transmission of information. Below that level, the genealogical hierarchy merges with the more general physical hierarchy of atoms and subatomic particles which is quite possibly infinite.

The upper limit of the genealogical hierarchy is bounded for a somewhat different reason. The largest-scale unit that constitutes a class of the genealogical hierarchy is the monophyletic taxon. *Monophyletic taxa* are here defined as aggregates of one or more species descended from a single common ancestral species. Thus, all life (hypothesized to be monophyletic due to the ubiquitous presence of RNA as a putative synapomorphy) constitutes a monophyletic taxon. So does the genus *Homo*. *The genealogical hierarchy is not the same as the Linnaean (taxic) hierarchy.* The genealogic hierarchy is a hierarchy of individual biological elements engaged in processes of information preservation, promulgation, and expression. The Linnaean hierarchy, in contrast, consists of a complexly nested array of taxa representing an aspect of the historical outcome of the evolutionary process.

Thus, monophyletic taxa are unranked in the genealogical hierarchy. The reason why monophyletic taxa belong in the genealogical hierarchy stems from their very nature: they are historical units formed by the ongoing production of new species from old. Such taxa constitute the highest, terminal level of the genealogical hierarchy simply because they themselves—these genealogically interconnected arrays of species—cannot by their very nature produce more of themselves, more entities of like kind. New species arise from old, a process we call speciation. Dobzhansky spoke of "raciation," a lower-level process of group reproduction. But we have no theories of "genusiation," "familiation," and so forth. Simpson (1961, p. 124) did in fact redefine monophyly in such a way that one could indeed speak of the derivation of one taxon of rank higher than species from another "of the same or lower rank" without violating such a redefined and broadened concept of monophyly. But such a concept, clearly related as it is to the overall evolutionary metaphor of the adaptive transformation of lineages, has remained little more than metaphor. We (especially paleontologists) speak of the origin of one genus from another, or of one family from another, but we have no hard and fast biological theory to explain how that might happen except simple recourse to a model of shifting allelic frequencies and the modification of the adaptive properties of organisms within a single (monospecific) lineage through a considerable segment of time. In other words, those who would claim that taxa of rank higher than species can and do make more of themselves at base have a purely reductive, transformational model in mind (or at heart). The concept itself is meaningless. Indeed, taxa construed in such a way are, once again, if not pure classes then at least the by now familiar classlike entities that cannot be construed as individuals in the sense used here. The genetic buck seems to stop with species. They are the largest-scale units of biological organization that, as a matter of course, engage in the process of making more individuals of like kind; hence all our models of speciation.

## On Individuality and More-Making

The discovery that the molecular biology of the gene is itself hierarchically structured gave a powerful boost to the general notion of a hierarchical arrangement of genetic information within the biological realm. Nonetheless, such molecular knowledge remains sufficiently embryonic (and my grasp of it sufficiently rudimentary) that I shall stick to the same definition of *gene* that G. C. Williams used in his 1966 book, *Adaptation and Natural Selection*. For my purposes, a gene is "that which segregates and recombines with appreciable frequency" (Williams 1966, p. 24).

When Williams (1966) and later Dawkins (1976, 1982) wrote of the (at least potential) immortality of genes, they were really speaking of the information that those genes hold. When I call a gene an individual, I do not refer to a locus that is responsible for coding some enzyme product. I refer to a particular segment of a particular chromosome within a particular cell of a particular organism. Such a gene is far from immortal. And for present purposes I speak only of germ-line cells, not somatic cells, all of which have copies of the genetic information copied from the germ line. It makes no difference, for the moment, that the gene for alcohol dehydrogenase is not used to produce that product while in a germ cell. Thus, while I agree with Williams and Dawkins that genetic information is potentially immortal (indeed, going back to Dobzhansky's lead, I believe that the conservation of such information provides a very large measure of what evolution is all about), on the other hand it is an incontrovertible fact that individuals within the genealogical hierarchy are at their most ephemeral at the level of the gene. The eggs of placental mammals may be retained intact throughout the reproductive age of the organism—in humans, forty years and more. Such cellular longevity is unusual; sperm is much more evanescent and ad hoc, as are gametes in general throughout the sexually reproducing eukaryotic realm.

Thus, in the very narrow, particularist sense in which I am using the word here, there is no problem seeing genes as individuals. Each is one particular segment of DNA within one particular germ-line cell. A gene has a corporeal existence. Though they are nothing like coherent, discrete beads on a string, we continue nonetheless to see the particulate nature of the behavior of genes, at least when we look at how they move around and how they operate within the soma. The array of base-pair sequences that superficially seems so continuous remains to us obviously particulate. Genes are indeed spatiotemporally bounded. They have births, deaths, and intervening histories, during which they may, in fact, change form.

The birth of such genes, of course, is their replication from a parent, anlage gene. And here, of course, Dawkins is perfectly correct: it is genes, and only genes, that in fact replicate. Replication is everything when we are dealing with the faithful transmission of information. Errors in replication are a major source of mutation, and here we run into a serious problem for evolutionary theory: Why does natural selection not remove replicative error? A favorite response is that selection somehow sees the need for the population-level variation, a construct justly troublesome to purists (such as G. C. Williams) who see so clearly that selection can have no eyes for the future. This sort of problem crops up over and over again up and down the genealogical hierarchy, and it is simply insolvable given the basic tools of logical analysis that the modern synthesis affords us.

But in stressing the replicative nature of the birth of new genic individuals, and thence contrasting such behavior with superficially similar birth processes of other entities, a very general point is usually lost: the production of new genes from old is just that. However faithful the copying process of birthing might be, we are dealing with a simple case of more-making, new genes from old. (Again, in terms of entities, I do not mean "change in form" when I say "new" and "old." Usually the new is an exact copy of the old.)

And such genes, even if we consider only those that form the entities of the germ line, do have histories. Such histories have figured only recently in our thinking of the evolutionary process—probably because while genes are simply sitting there as segments of DNA strands we are unaccustomed to thinking about them doing anything at all. Cells that are not dividing are in a "resting stage." Yet the new molecular biology of the gene has revealed a world of semi-autonomous activity in the genome, where transposable elements can move to different positions within and among chromosomes, to the point, apparently, of biasing the replication of some alleles over others. Quite a bit can happen to the particular segment of DNA that we call a gene during the course of its "life." Mutations in the broadest sense can occur at any time, during birth, as mistakes in copying, but also as spontaneous changes at any stage during the course of a gene's existence.

And genes die. Their reproductive proclivities and faithful replication may lend them the aura of immortality, but genes as individuals are ephemeral. Normally, the deaths of individual genic entities are of no consequence. Their clones are there to carry on. But among allelic forms, differential survival biases the genetic information that lives on. And the deaths of genes can also be important to the evolutionary process if it can be shown that there is a differential survival of gametes with some particular set of alleles. Meiotic drive and related phenomena are cases in point.

Chromosomes are another matter. Dobzhansky (1937a, p. 74) actually called chromosomes "individuals." Eldredge and Salthe (1984) pointed to the difficulties inherent in including chromosomes in the genealogical hierarchy, deciding in light of the objections to omit them entirely. Chromosomes, for one thing, are not ubiquitous in nature; they are confined to the Eukaryota. But that objection loses its force when we realize that the genealogical hierarchy exists even when one or more of its constituent elements might be missing as a sort of special case. Indeed, I have already so argued in the case of demes; asexual organisms are not arranged into species (or demes, for that matter), yet they have genes, chromosomes (many do, at least), organisms, and clones. And it is just as clear that there is evolution within such a system. My position insofar as chromosomes are concerned is that when thinking of prokaryotes we simply do not have a chromosome level in the genealogical hierarchy. When we deal with eukaryotes, however, it might very well be profitable to include such a level in the process hierarchy of genealogy.

Chromosomes are clearly discrete individuals, in a direct corporeal sense, but only during some phases of cellular history, specifically when the cell is undergoing subdivision. Yet this objection loses force if it can be shown that processes important to the transmission of genetic information occur as a result of that information being packaged into discrete chromosomes. And this is manifestly the case. Dobzhansky (1937a) devoted an entire chapter to chromosome-level changes. Within- and among-chromosome maneuvering of genetic elements have by now become familiar, and the inversions, deletions, crossings over, and so forth from cytogenetics are as clearly important today as they were to Dobzhansky in the 1930s. Much goes on of evolutionary significance at the chromosome level in organisms that have chromosomes.

In what sense do chromosomes reproduce, at least independently from the base-pair sequences of the DNA molecules? For if chromosome reproduction is straightforwardly and entirely reducible to DNA replication, then we can perhaps argue that the chromosome level is nothing unique and validly to be distinguished from the gene level within the genealogical hierarchy. Yet chromosomes are very much more than double-helix strands of DNA; the proteins of chromatin and indeed the entire intricate physical structure and arrangement of chromosomes make them far more than simple linear genic strands. Chromosomes may assemble and disassemble, but there really can be no doubt that when they do occur they are (in particular instances) true individuals, historical entities with births, histories and deaths.

Chromosomes are nearly as ephemeral as genes. To the extent that genic and chromosomal death is a function of a higher level cellular or even organismic death, the two are equally ephemeral. However, it is at least conceivable (if perhaps of no major significance) that intracellular processes can cause the death of a particular genic individual while the chromosome on which it sits lives on. And I have already briefly alluded to the birthing process of chromosomes; it is important, I think, that many more kinds of gross rearrangements of the genetic material in general can occur with respect to the duplication of entire chromosomes than the simple replication of the triplicate base pairs (codons.) These "chromosomal mutations," as Dobzhansky called them, are of an entirely different order of magnitude. Finally, even during the lifetime of chromosomes, events can occur that can have profound influences on the transmission of information. For example, Down's syndrome may arise in some cases because chromosomes 21 and 15 lie in close conjunction for years in the egg cell and in some cases do not dissociate when meiotic division eventually continues. The message seems clear enough: where they occur, to incorporate those processes that affect information continuity, chromosomes, rather like demes, simply must be included in the genealogical hierarchy.

Cells, too, pose a problem, if only because they obviously make more of themselves. And germ-line cells, with their patent discreteness, their births and deaths and histories, do indeed belong in the genealogical hierarchy. Cases of "sperm death" have been documented (arising from lower-level, intracellular processes). In general, however, relatively little attention has been paid to the cellular-level aspects of continuity (transmission) of genetic information from an evolutionary (as opposed to a purely physiological) point of view.

Somatic cells, however, belong to a hierarchy of organisms and are most regularly and appropriately considered in terms of the ecological hierarchy—in other words, matter–energy transfer rather than information flow. The exception to this generalization, however, is in the unicellular organisms, where the entire organism *is* a cell (Hyman's "acellular" organism to the contrary notwithstanding). Organisms, because they make more of themselves and because they obviously transmit genetic information, must belong to the genealogical hierarchy, in spite of the fact that they also belong to the ecological hierarchy.

And, of course, organisms are the paradigmatic individuals, so much so that it seems difficult, as well as perhaps counterintuitive, to see other sorts of entities (especially those of a higher level) as themselves constituting individuals. The significance of organisms in the evolutionary process, as generally conceived, lies in the expression of their genes. The organismic phenotype is the locus of most adaptations—for all intents and purposes, we might say all adaptations. It is the nexus of environment and information. It is the locus of (natural) selection. There is every good reason why the modern synthesis (and evolutionary theory in general from Darwin to Dawkins) has focused on this important nexus. [As Dawkins himself stresses (especially 1982), his view of the selfish gene is more of an emphasis on one side of a Necker cube, where the same structure flips back and forth from one perspective to another, meaning, in his case, from seeing organisms as the central object, their genes merely the ledger keeping track of phenotypic change, to his preferred version of seeing organisms as merely the external expression of the all-important genes.] Organisms retain this special position in the hierarchical structure that I advocate here; organisms lie in both hierarchies.

The role conventionally seen for organisms in evolution—as the carriers of adaptations—has more to do with their status as members of the ecological hierarchy than as members of the genealogical hierarchy. Most adaptations fall under the role of organisms as interactors (in the sense of Hull 1980), but some are for reproduction. As is well understood, sexual organisms possess specific behavioral, physiological, and anatomical properties that are designed for the purpose of successful mating with one or more other organisms. It is solely this aspect of organisms that concerns us here: organisms reproduce, they make more of themselves (whether sexually or not), and they are thereby essential links in the maintenance of genetic continuity.

Here at the organismic level somewhere around the middle of the genealogical hierarchy, there is a transition between replicative faithfulness and longevity. Asexual organisms reproduce, and the limit to their replicative accuracy comes from the cytogenetic and molecular limits on accuracy—purely lower-level constraints. Sexually reproducing organisms, as Darwin was well aware, are not replicators. In general, offspring resemble parents—enough of a match for the notion of selection to work. The very existence of among-organism variation, like mutation, can be seen as a problem in the weltanschauung of the modern synthesis. The usual retort is a shrug. If we cannot explain why selection hasn't removed all this variation, we should at least be grateful that it exists, else there would be no raw material for evolution to work on, no escape clause against the certain, if unpredictable, change in the environment. And there is something appealing to this line of thought, if it is only that it addresses the question of how organisms have managed to last more than three and a half billion years on earth.

Among-organism variation is, of course, a demic consideration. I broach it here only because the unspoken assumption in evolutionary theory seems to be that (sexual) organisms are imperfect replicators. They are imperfect replicators, of course, but the attendant implication is that they shouldn't be, that there must be some benefit to organisms derived from recombination that overrides the importance or desirability of perfect replication—as if reproduction benefits an organism, selfish genes notwithstanding! Again, I pursue these points in greater detail in my next chapter. Here I simply stress that organisms, asexual and sexual, regularly and routinely participate in the process of making more organisms, more "individuals of like kind." Degree of replication declines, but duration increases. The packaging that houses the genetic information has a life span typically greater than that of its sex cells, chromosomes, and genes. Moreover, if an organism reproduces at all, it may do so more than once, each time producing another organism (or batches of them) with the same attributes of longevity and reproductive potential. And we should always remember Darwin's dictum that organisms do indeed resemble their parents.

There is no point in belaboring the obvious: all organisms have births, histories, and deaths. As quintessential individuals, they are clearly spatiotemporally bounded. Only a few additional comments need be added. Some organisms undergo a series of one or more radical transitions—metamorphoses—in the course of their existence. Holometabolous insects are such creatures. From the standpoint of the genealogical hierarchy, the ontogenetic stages, however discrete, of an organism all belong to a single individual; the larva, pupa, and imago of a single organism of a butterfly species is all one genealogical individual. From an ecological point of view, larva, pupa, and imago might arguably be considered to constitute three separate individuals, with radically different roles to play in the economy of nature. But genealogically—genetically—they are but one.

Colonial organisms likewise pose problems of individuality, but again more from an ecological, economic point of view than from a genealogical perspective. Colonies of attached, undifferentiated coelenterate polyps, where each polyp performs all normal physiological and reproductive functions, clearly hold two levels of individuals. The polyps and the colony as a whole are both individuals, and neither is an organism in precisely the same sense as Charles Robert Darwin is an individual. Differentiated colonies are another matter; each organism in a social bee colony is, of course, an organism regardless of whether it reproduces or not. The entire colony is a functional and genetic whole, thus also very much an individual. It could be argued, for instance, that workers as organisms belong only to the ecological hierarchy, their functions limited strictly to the economic sector and their contributions to the fitness of the queen notwithstanding. They themselves do not reproduce. But it could as well be argued—as indeed it usually is—that workers are very much involved in activities that can be construed as reproductive. For present purposes I am content merely to establish that organisms belong to the genealogical hierarchy. Common-sense observation of nature reveals far more of the economic roles organisms play, and their general failure to replicate precisely has devalued their status insofar as genealogy is concerned—to the point where my colleague S. N. Salthe (1985) has removed them entirely from the genealogical hierarchy.

One final word concerning organisms in the genealogical hierarchy:

Weismann's doctrine has withstood its latest challenges (Steele 1979) and the received truth remains: whatever happens to an organism during its lifetime, its history including its birth and (obviously) its death, it will not transmit the results of such events to its offspring. This commonplace item of received biological wisdom marks a transition in the genealogical hierarchy: hitherto, at the levels of the gene, chromosome, and sex cell (gamete), what happened during the lifetime of the individual could very much affect the maintenance and transmission *as well as the nature of* genetic information. What happens during the history of an organism will easily affect whether that organism reproduces (in itself no small consideration), but it may no longer affect the nature of the genetic information unless the germ line itself is affected (as by exposure to environmental mutagens).

We move on to demes. Demes are local aggregates of interbreeding organisms; populations are localized aggregates of conspecific interacting organisms. The two are easily confused, and the words (plus other synonyms, such as Dobzhansky's use of the term *colonies*) are often used interchangeably. In point of fact, the fuzziness of the very concept of population, not to mention deme, is notorious. For most purposes, a deme consists of the breeding (reproducing) members of a population. The key element of a population is interaction, mostly concerned with energy exchange; of a deme, it is reproduction. Otherwise, the two may well be virtually coextensive. Rather like the dual economic and reproductive aspects of organisms, demes and populations (though not identical) represent the dual functions of nearly coextensive local aggregates of organisms.

What makes a deme a deme? If they are historical entities, what makes them coherent, what conveys their continuity, imparts their boundedness? Multicellular organisms are held together by cellular cohesion and some form of coherent wrapper; moreover, they depend upon continued cell division, and, at base, continued somatic cell division, for their very continuity. Purely reproductively speaking, organisms depend upon the (usually) continued production of gametes. When we reach the upper three levels—demes, species, monophyletic taxa—all integrity is imparted by a simple more-making of the next lower individuals. Some organisms reproduce, making more organisms, some of which in turn reproduce, and the deme is simultaneously defined (bounded) at any one moment and given its temporal continuity. Organisms are not immortal, but demes potentially are.

Neither populations nor demes, of course, are bounded by iron walls. Some species appear not to be easily segregated into such units. On the other hand, some of the organisms inhabiting the kopjes that jut out from the grassy plains of eastern and southern Africa are alleged to be restricted to one particular individual kopje. Rock and bush hyrax and klipspringer, for example, are apparently so restricted, while giraffe, leopard, and lion freely range among kopjes. A klipspringer (*Oreotragus oreo-*

*tragus*) deme is therefore a clearcut spatiotemporal entity, virtually coextensive with a klipspringer population (which, however, includes all living klipspringers on a kopje at any given moment, including those of pre- and postmating age and those of reproductive age that, for some reason or another, do not breed). No species is known to blanket, uniformly and absolutely, the entire area within its overall range. Thus, all species to some extent really must be fragmented into demes.

And demes, it appears, do come and go. If they last longer than the lifetime of a single organism, they are subject on a broader scale to the vicissitudes of life. Demes, specifically, cease to exist almost solely when and if a local population ceases to exist, for purely economic reasons. Put another way, demes cease to exist for reasons that almost never have anything to do with what makes them demes, their demal properties, which boil down to the reproductive behavior of organisms. Reproduction ceases as a byproduct of a more fundamental challenge to the economic existence of organisms, hence a threat to the entire population. In particular, it is commonly appreciated that of all the major classes of physiological activity in which organisms engage, reproduction is the only one not essential to an organism's survival. The predicted corollary is typically observed in experimental situations: when stressed, as when organisms are placed in physicochemical milieus at or near their known tolerance limits, reproduction is the very first activity that ceases (see, for example, Bradshaw 1961, on foraminifera).

Similarly, demes are incidentally founded (born) as a mere side effect of the founding of new ecological populations, a process of colonization that reflects simple habitat availability, ability on the part of organisms to get to the habitat, and consequent exploitation of the newly available resources.

But demes, despite some of their almost intuitively obvious properties, remain difficult to characterize exactly or to study in the field with any degree of precision. Indeed, they are probably best studied by mathematical modeling, where their numbers and boundaries are set by fiat. Demes remain critical if we are to utilize Wright's "shifting balance theory" and all its implications. But Dobzhansky (1937a) was evidently correct when he pointed to particulate genes and species as the two main sources of discontinuity relevant to evolution. For, ironically enough, the next higher level in the genealogical hierarchy is *less* diffuse than the deme. That, the level of the species, forms one of what in Simon's (1962) terminology is a class of "stable forms"—stable entities—within the genealogical hierarchy.

## Species

Are species real? A brief summary of the history of the problem would go something like this: The pre-Darwinian majority saw species as both real and fixed. William Whewell (1837, p. 626) perhaps said it best: "Species have a real existence in nature, and a transition from one to another does not exist." Darwin, seeing that characterization of species as antithetical to the very notion of evolution, developed a picture of adaptive fluidity in which species are seen as classlike stages in a passing stream of anatomical transformation. The synthesis, especially through the eyes of Dobzhansky, Mayr, and Simpson, tended to concur when it came to seeing species temporally and when the larger patterns of anatomical change were considered. But Dobzhansky and especially Mayr tended also to see species as discrete entities, much like *Paramecium* individuals, to return to Mayr's (1942, p. 152) explicit imagery. There is now on the floor the further proposal that species are individuals temporally as well as spatially—a suggestion we owe more to Ghiselin (1974a) than to anyone else. And I have already suggested (chapter 5) that the data of paleontology (and, I would add, the comparative systematics of monophyletic genera) amply support such a position.

So much for history. Is there any substance to the view that species are not mere arbitrary abstractions, classlike entities of no particular importance to evolution? Are they really ephemera, as Dawkins sees them, or, worse yet, arbitrary figments, as Levinton (e.g., 1983) persists in seeing them? Or are they rigid monoliths of morphological conservatism, as champions of punctuated equilibria are occasionally prone to depict them?

Dobzhansky (1937a, 1941) and Mayr (1942), drawing on their forerunners, defined species as groups of sexually reproducing organisms, groups defined by the ability to interbreed among themselves and not with members of other such groups. "Groups" smack of "sets" (which are classes), but the point is clear enough. The contention is simply and solely that among sexually reproducing organisms there are groups that are defined and recognized simply because they exist as such reproductive plexuses. That nature is in fact organized in just this way seems patently obvious to me.

And here arises a superficially serious case against the biological species concept and an apparently illogical construct of the genealogical hierarchy. It may be supposed that, if demes are composed of a subset of reproducing members of a local economic population (itself a rather classlike concept), then species logically should consist only of those organisms that are interbreeding at any one moment—if we define species as reproductive communities. (Perhaps this is the reason why Mayr originally included the phrase "actually or *potentially* interbreeding" in his definition.) Doesn't the informational aspect of the genealogical hierarchy force us to this patently absurd conclusion?

No, it doesn't. Despite their classlike overtones, the biological species definitions of Mayr and Dobzhansky were geared to describe coherent entities, their problems in extending the concept temporally notwithstanding. There are two retorts to the objection; one lies within the (inappropriate) classlike conception of species, and the other is in a context of seeing species as spatiotemporally bounded entities—individuals. Thus, we could counter the objection by defining species as all organisms possessing a particular collection of reproductive adaptations—Paterson's SMRS (species mate recognition systems)—regardless of when, if ever, they put them to use. From a classlike standpoint, the response makes sense. But the real answer to the objection is that all organisms—hence all demes—within a species are produced by the reproductive activities of other members of that species. That some organisms of any given generation do not reproduce is important for stasis and change in genetic information but irrelevant for any definition of species.

That reproductive plexuses persist for more than one generation is also known to us all. That they may persist for many thousands of generations can only be derived as an inference. After all, all sexually reproducing organisms appear to be interrelated (as is indeed all of life); thus, there must be a skein of such reproduction since the origin of sexuality. The empirical support that the punctuated equilibria concept truly adds to the argument that species are spatiotemporally discrete is as follows: (1) We see different anatomical packages at a localized "instant" in geological time. (2) We sample virtually identical packages at different horizons, younger and older than our initial observation. (3) We infer that we are dealing with an ancestral–descendant skein, an inference that clearly may be wrong in any particular instance but that, if correct in some of the many putative instances, leads us ineluctably to conclude that species have a nontrivial distribution in time. The argument does not hinge on stasis, the observation that, once evolved, species tend to change relatively little. Stasis merely makes the identification of species through time a little easier.

The actual argument, then, that species are spatiotemporally bounded reproductive plexuses is precisely that: an argument of necessity. There are these coherent reproductive plexuses in the world today, and it is no startling inference that sexual reproduction has been going on for a long time (more than a billion years, in fact). But what accounts for this packaging? Why are species discrete (to repeat Dobzhansky's query)? If mate recognition provides the *spatial* boundedness, what are the *temporal* bounds of a species?

Species really have been the most confused and confusing entities, at least so far, in the history of evolutionary thought. I sincerely hope that we can now break away at last from most of that confusion. To this day the notion of reproductive community is still confounded with the notion of phenotypic attributes of component organisms. The entire gradal notion of species (stages of differentiation along a lineage) sees species as classlike clusters of organisms that share a certain suite of adaptive phenotypic characteristics. Eldredge and Cracraft (1980) saw the distinction between these grossly different classes of attributes (both adaptations, one for reproduction and the other economic in nature) yet retained the two in their definition of species because systematists (usually) and paleontologists (nearly always) have no information about reproductive behavior. Operationally, we use morphology to determine species. But epistemologically derived yardsticks for the identification of species need not afford accurate definitions of what species actually are. They are reproductive communities first and foremost, communities in which the component organisms need to be different from their nearest phylogenetic neighbors only in some minimal way pertaining to their reproductive behavior.

The work of H. E. H. Paterson (1978, 1980, 1981, 1982) and colleagues is enormously helpful in this regard. I have already alluded to Paterson's notion of the specific mate recognition system (SMRS). Paterson sees all isolating mechanisms as really ways (behavioral, chemical, physical) of recognizing appropriate—conspecific—mates. Nor is the point trivial: according to Paterson's analyses, there is a fundamental misconception inherent in the very term *isolating mechanism*, for it implies (as we have seen, especially in the lexicon of its inventor, Dobzhansky) that these phenotypic suites serve the expressly adaptive purpose of the prevention of interbreeding between species. It is true that both Dobzhansky (1937a, 1941) and Mayr (1942) did occasionally look upon so-called isolating mechanisms as positive features to ensure conspecific mating. But if we follow Paterson's lead and look at his SMRS only this way, much of the confusion that has attended the entire notion of species and speciation falls by the wayside. For one thing, Paterson's perspective makes us see the isolation between species as an accidental byproduct; selection is for mate recognition, not for reproductive isolation. Mayr implicitly saw the point, while Dobzhansky basically did not.

Speciation, then, is an accident. Selection for continued mate recognition in situations where allopatric populations are isolated may (or, of course, may not) lead to changes in mate recognition systems such that organisms from parental and daughter species may meet but not mate in neosympatry. Species bud off from one another for no good reason other than the need of organisms to continue to recognize mates. And the origin of species—their lower boundedness, their beginnings as discrete entities—arises as an accident of geography and the propensity, even the need, of sexual organisms to continue mating.

It has long been a theme in evolutionary biology—as we have seen plainly in Mayr (1942)—not to see species as "adaptive devices" (to use Paterson's term). Paterson (e.g., 1981, 1982) recognizes that organisms within species are adapted to their habitats. Adaptation in an economic sense presumably takes place via natural selection on a generation-by-generation basis within demes, at least potentially. But adaptive divergence in this economic sense alone will not lead ineluctably to new species; only change in that particular subset of phenotypic characteristics that constitutes the SMRS will lead to the formation of new species. Adaptive change in the economic sense seems more a response to the opportunities afforded by embryonic reproductive isolation, an opportunistic reaction, as Carson (e.g., 1982) has discussed. The usual view, certainly common in the synthesis, looks at speciation in just the reverse way: speciation seems to be more a function of such (economic) adaptive change, both from the standpoint of the development of reproductive isolation and from the perspective of speciation as simply the slow accumulation of adaptive change through time, the simple transformation of phenotypic properties within an evolving lineage. The theory of punctuated equilibria is certainly relevant here, with its central tenet that most morphological change within the history of monophyletic taxa seems to take place in conjunction with speciation events as opposed to during the relatively far longer histories of species once established. At any rate, species can become internally adaptively differentiated to a rather great extent without reproductive isolation fragmenting the species into two or more separate species—just the conclusion reached by Mayr (1942), even though speciation was in places argued to be the accumulation of such economic adaptive differentiation, where the adaptations start to function secondarily as isolating mechanisms. What Paterson's scheme implies is a clearer understanding of the difference between reproductive and economic adaptations. The former are devices to promote the more-making activities of organisms. The latter are devices to ensure somatic growth and maintenance, which may, of course, have implications for the reproductive success of those organisms.

It is apples and oranges. We tend to see species as aggregates of similarly adapted (economically) organisms. Yet we also see them as reproductive communities. We have the old (Mayr, Dobzhansky) twin themes of diversity versus discontinuity in slightly different guise. Species clearly cannot be both sorts of entities. In terms of framing a coherent definition of what a species, *any* species, is, either we must see species primarily as collections of similarly adapted organisms in an ecological sense (albeit those similar organisms can mate and reproduce), or we must see them primarily as reproductive communities (in some of which the constituent organisms may look pretty much alike, while in others there may be a great range of adaptively based variation).

To see species as clusters of similarly adapted organisms is to see them as classlike entities with no necessary "bridgeless gaps" between them and their next nearest phylogenetic neighbors. Indeed, saltation has never proven to be a particularly successful or appealing model of adaptive phenotypic change (largely for want of supportive evidence rather than through lack of imagination); the expectation would be, I believe, that the organic realm would be more intergradational (phenotypically speaking) than what we see before us—that is, if species were merely clusters of similarly economically adapted organisms. Indeed, Dobzhansky and Mayr saw the morphological gaps between species and tried to explain them, Dobzhansky with reference to isolated habitat-characterized adaptive peaks, Mayr perhaps more as outcomes of reproductive isolation.

To see species, on the other hand, primarily as reproductive communities is to acknowledge a basic aspect of these clusters of sexual organisms, which really are reproductively isolated from other such groups, and to accede to a ready explanation for why there are such groups—simple selection for continued mate recognition in allopatry. This amounts to saying that species are stable entities (or "stable forms"—Simon 1962, with explicit reference to hierarchy theory), formed automatically as a consequence of sexual reproduction. I must follow Paterson here and adopt that portion of the arguments of the synthesis that saw species purely as reproductive communities, communities of interbreeding organisms, discrete from other such groups (hence spatially bounded) and with discrete beginnings (speciation) and ending (species extinction). Species seem to be spatiotemporally bounded individuals, but only if we think of them as reproductive communities.

This sets the stage for an important argument and something of an irony. The struggle to return to a notion of species as spatiotemporally bounded entities—as individuals—seems to have been won. And species conceived of in this fashion patently belong in the genealogical hierarchy; they are individuals bound together by the ongoing reproductive activities of organisms (where there are no obvious demes)—or of the continued existence and production of demes where they do exist. But arguing that species should be included as individuals within the genealogical hierarchy by no means automatically excludes them from consideration as members of the ecological hierarchy. Yet they cannot be—and therein lies the irony.

When Ghiselin (1974a, 1974b) first fashioned a coherent argument that species should be construed as individuals, he drew an explicit analogy between such biological entities and economic entities; thus, according to Ghiselin, *Homo sapiens* within the general class of species is analogous to Gulf & Western within the class of international conglomerates. The analogy is apt, up to a point: both are individual entities. But the analogy has been pushed perhaps too far. As Hull (1980) found, species simply are not interactors. Just the fact that species are historical entities—as are, say, organisms—does not establish that they have a role in the economy of nature or what such a role might be. The kinds of dynamic processes generally envisaged for entities in the ecological hierarchy generally involve energy exchange or other forms of direct interaction. It is possible, obviously, to see organisms in this light. Populations (of conspecifics) also play definable roles within communities, which are cross-genealogical arrays of populations living in the same habitat. Communities, moreover, are integrated into larger structures along the same lines.

How do species fit in, in this particular context? Apparently, they simply don't. Species have a role to play in the ecological theater, but I cannot see it as the holistic role of active, integrated members of some dynamic system. Unless a species is restricted to a single population, it is impossible to go to nature, sample a species, and see it as a dynamic interactor in the ecological realm. Thus, from the economic point of view, species do seem too diffuse to play a particular role, to serve as a single, unified, coherent entity in an energetics sense. To my knowledge, no ecologist has adduced a theoretical structure that sees entire species integrated into some ecological system. In a purely ecological, economic, energetics sense, species do not exist.

Local populations of virtually all species (extant and fossil) known to me are integrated into a variety of local ecological situations. In the simple example of resource bases utilized by such local populations, there is usually greater variation among rather than within such local populations. Each population (i.e., of conspecifics living allopatrically) plays a slightly different role and indeed is integrated into a slightly (or even conspicuously) different ecological setting from those of the next population. Thus, a species can play an economic role in the additive sense of its component populations and, of course, of its component organisms. (I take the position immediately below that populations do play a role in an economic sense over and above their component organisms in their integration into local communities or ecosystems.) This is to say, for example, that the species *Panthera pardus* (leopards) is a far-flung, disjointed, yet apparently functionally coherent interbreeding system (a reservoir of information, both economic and reproductive), yet is not meaningfully to be construed as a single entity in some supposed economic system imagined to run from southern Africa over to Indonesia. Leopard prey items vary tremendously over this huge species range. In any case, in ecological terms, the area defined by the present (or any past or future) range of this species is difficult to construe in terms of a single, coherent economic entity.

Finally, if it is true that species are not meaningfully to be considered as economic individuals, the argument must hold with more force for monophyletic taxa of rank higher than species. The idea that Felidae occupy a single, coherent range of closely associated peaks on some large-scale adaptive landscape—when taken to mean that Felidae play a specifiable, coherent role in some large-scale terrestrial biome—amounts to empty metaphor, furnishing an inaccurate description of nature.

I find irony here. A major trend, certainly in paleontological circles, has been to acknowledge the individuality of species, to see species empirically coming and going in the records of systematics, biogeography, and paleontology, and hence to see them as genuine actors in the evolutionary drama. And indeed they are. But the assumption—at least the assumption behind all early notions of species selection—is that species are active in the economic sense as well. Note that this is not precisely the same issue as that encountered in chapter 5, that species really do not seem to have emergent properties, adaptations in the economic sense beyond those of their constituent organisms. Certainly most early claims for species selection involved such structures of the phenotype, economic adaptations (such as brain size in fossil and recent hominids) that are plainly visible as the properties of organisms. [Even the claim (e.g., Arnold and Fristrup 1982; Gould 1982b) that differential rates of speciation might arise as species selection from different species-level reproductive modes for the most part seems a matter of the behaviors, physiologies, and anatomies of organisms]. The case for species selection seems weak on these grounds alone. But in acknowledging that species are best construed as reproductive communities, and that, in any case, there is no evidence that species *as units* play a direct role (or roles) as interactors in the exchange of energy in the economy of nature, we remove them wholly from the realm of selection in the economic sense.

Species, then, do exist. They are real. They have beginnings, histories, and endings. They are not merely morphological abstractions, classes, or at best classlike entities. Species are profoundly real in a genealogical sense, arising as they do as a straightforward effect of sexual reproduction. Yet they play no direct, special role in the economy of nature.

A final note to complete the formal picture of species as members of the genealogical hierarchy: a species is terminally bounded—ceases to exist—when the ongoing among-deme (or simply among-organism) reproductive skein is itself terminated. Thus, "extinction by transformation" is meaningless. Extinction through evolution into something else can only happen in the species-as-classlike-entities sense, a definitional artifact of seeing species merely as arbitrary segments along a gradient of phenotypic transformation. Such unbroken skeins, regardless of the degree of morphological differentiation accrued, must be construed as single species (Wiley 1978; Eldredge and Cracraft 1980—not to mention the very definition of "evolutionary species" in Simpson 1961, p. 153).

And species have births. Again, such births must be speciation in the sense of the development of one or more descendant reproductive communities from an ancestral reproductive community. The appropriate place for speciation theory within evolutionary theory is in the context of simple more-making of species. And that more-making is precisely what provides the glue that holds together the final, highest-level entities of the genealogical hierarchy—monophyletic taxa.

## Monophyletic Taxa

If it is difficult to see species as individuals, as discrete historical entities, the situation is worse for taxa of higher categorical rank. Yet placental mammals clearly had a beginning, have had a history, and will inevitably have a termination at some point in the future. Such taxa—pure genealogical strings of species, such as the placental mammals, birds, Eukaryota—are marked by the "invention" of evolutionary novelties, presumably adaptive aspects of the phenotypes of organisms that serve as devices either for economic interaction (and indirectly facilitating reproduction) or for reproduction directly. The complexly nested hierarchy of such taxa is thus aligned with the matrix of hierarchical systems that such transformed novelties (homologies—cladists call them synapomorphies) tend to form.

Yet it is not true that the hierarchies of homology are the same as the Linnaean hierarchy (see Chapter 7 for additional discussion on these two "resultant" or "pattern" hierarchies). Such a confusion can only arise if taxa are classes. The properties by which we recognize any organism as being a mammal—items such as hair, three middle ear bones, mammary glands, placenta, and so forth—are putative evolutionary novelties inherited by all descendants of the "Ur-mammal." But these characters are *not* "Mammalia." "Mammalia" is not "hair." Mammalia is a finite number of species that have descended from a common ancestral species whose component organisms first possessed those certain properties, such as hair, three middle ear bones, and so forth.[2]

Thus, on an abstract level, such taxa seem to be classes: segments of the biota defined and recognized by one or more phenotypic features of organisms. If, on the other hand, we consider the peaks and ranges of Dobzhansky's imagery for the adaptive history of life, we see higher taxa not purely as classes nor purely as individuals, either. But remembering Dobzhansky's observation that the two basic sources of discontinuity in evolution are genes and species, it is no great conceptual leap to see such ancestral–descendant arrays of species as themselves constituting definite entities, clearly spatiotemporally bounded. And, almost as a matter of definition, it is the continued production of new descendant species that keeps a monophyletic taxon going. All of the species of Bivalvia in the Ordovician are now extinct. On the other hand, all the species of trilobites in the Ordovician are extinct as well, as are all species of trilobites that have ever existed; the trilobites as a large taxon are extinct. Monophyletic taxa of rank higher than species have characteristically longer durations than their component species—a fact of no small consequence in our description and interpretation of the history of life. They are, on average, more resistant to extinction than species—one respect in which relative ranking in the Linnaean hierarchy may make a difference in evolutionary processes.

The reason why the Linnaean hierarchy is not simply tacked onto the top of the genealogical hierarchy is that there is no formal difference between genera and kingdoms. All monophyletic taxa are defined as "all species (one or more) descended from a single ancestral species." I have also repeated the common observation that there is really nothing about the process of the origin of new genera, say, that is different from the process involved in the origin of new families, orders, classes, and phyla. Only if taxa are classes (in the philosophical, not the Linnaean, sense)— sets of organisms possessing some anatomical novelty judged to be sufficiently great—do we get into a mode of argument that visualizes the origin of higher taxa as perhaps involving the radical restructuring of bauplans, in contrast to the far more modest phenotypic change typically accompanying speciation. Indeed, as we have seen, this seems to have been precisely the way in which Simpson viewed micro-, macro-, and megaevolution in 1944. I am not saying that the characterization—and imagination of radical transformation—of bauplans is of no material significance to our consideration of the history of life. But the appropriate level of such theory and investigation is in the developmental biology of organisms, not at the supposed level of taxa of higher categorical rank.

However, inasmuch as there does tend to be a correlation between the rank of a taxon and the relative number of its included species, there may be some obvious corollaries in terms of the nature of the histories and deaths of monophyletic taxa dependent upon such rank. But such correlations are invariably sloppy. A new phylum, the Loricifera, recognized in 1983, is based on relatively few species. The phylum is recognized simply because the comparative anatomy of its included organisms so far studied reveals no clearcut pattern of synapomorphy with any as yet defined and recognized phylum of animals. Thus, in asserting that monophyletic taxa are not classes, nor merely classlike entities, but full-fledged historical entities—spatiotemporally bounded individuals—I am not denying that there may be some systematic differences in the evolutionary significance of such taxa dependent upon their rank. But as a first-order approximation—and especially because monophyletic taxa do not beget monophyletic taxa (i.e., they do not make more of themselves)—I see such entities as the single-level upper terminus of the entire genealogical hierarchy.

## THE ECOLOGICAL HIERARCHY

The genealogical hierarchy seems, somehow, to make sense, to appeal to our notions of what is properly to be construed as evolutionary in biology. If the ontology of some of its components needs the sort of revamping that Hull (1980) called for, nonetheless the names are familiar; from genes to higher taxa, I have introduced no terms novel to the discourse. All these entities, in one guise or another, figure in the writings of the modern synthesis.

Yet when we peer out the window at nature we see none of these entities of the genealogical hierarchy, save organisms, of course. What we see instead is an eastern gray squirrel with an acorn in its mouth, darting across a lawn, worried about a neighborhood cat, scrambling to safety, perching on the lower branch of a white pine, there to manipulate the acorn, shell it, and eat it. Our senses seem to be telling us that on a moment-by-moment basis nature is spatially and functionally organized in a manner very different from the pattern suggested by the genealogical hierarchy. It is in this arena that the actual activities of life are played.

The ecological literature reveals a rather startling diversity of definitions for such familar terms as ecosystem and community—so much so that Hull's call for a clarified ontology transcending the common-sensical seems equally applicable to such ecological units. Some ecologists, for example, take strong issue with the suggestion that communities can be construed as individuals. The problem seems to come from the apparent lack of definitive boundedness to such entities. Whereas a certain amount of spatial restriction might be imparted (if through nothing other than the distribution of a physicochemical regime, such as rate of water flow, oxygen tension, temperature, and pH of a mountain stream), communities seem to be constantly changing, seldom the same as sampled from moment to moment.

Indeed, ecological entities above the level of the organism seem primarily spatially construed, whereas it is the historical nature of species and especially of monophyletic taxa that often seems more significant to us than their spatial distributions. Demes are sometimes claimed by ecologists to be subunits of populations (as in the usage of MacMahon et al. 1978), but in the classification I adopt here demes are strictly genealogical entities. In any case, demes are generally treated more from a spatial than a temporal (historical) point of view. Remember at this juncture that all such entities, if they really are individuals, are both spatially bounded and historical entites, even if their temporal continuity is rather brief (in human terms). Epistemological problems, particularly those induced by a consideration of scale, are rife in all considerations of hierarchy.

If various sorts of ecological entities are difficult to see as individuals, ecologists on the other hand have had no trouble describing the natural world in hierarchical terms. The metaphor is common in textbook treatments, and Valentine (1968, 1969, 1973) Salthe (1975, 1984), Allen and Starr (1982), and MacMahon and colleagues (MacMahon et al. 1978 1981; Fowler and MacMahon 1982) have provided recent explicit treatments of ecology from a hierarchical perspective. Indeed, Valentine's 1973 book, *Evolutionary Paleoecology of the Marine Biosphere* (in my opinion vastly underrated) is entirely organized around the hierarchical organization of ecological units and contains much food for thought on the interconnections among ecological systems in particular and evolutionary systems in general. Allen and Starr (1982) have also produced an entire book (*Hierarchy: Perspectives for Ecological Complexity*), but in eschewing ontology they have, I believe, seriously limited the utility of their thought.

More useful, at least to someone frankly groping for an ontology of ecological entities, is the work of Valentine and of MacMahon and colleagues. Valentine's classification (table. 6.1) of ecological entities entails individuals (organisms), populations, communities, provinces, and, finally, the entire (marine) biota. Their respective equivalents in functional terms are the functional range of individuals, niches, ecosystems (community systems), provincial systems, and, finally, the marine biosphere. Valentine's dual system of terminology reflects the usual distinction in ecology between levels of organismic association on the one hand and matters of energetics, which automatically involve the abiotic realm, on the other. As Eldredge and Salthe (1984) have discussed, one achievement of paleoecology in recent years is the demonstration that recurrent assemblages of commonly preserved organisms may typically persist for millions of years. In other words, organisms from a number of different species

Table 6.1   Valentine's classification of ecological units

| Genetic term | | Ecological term | |
|---|---|---|---|
| Unit of organization | Organization | Functional | Descriptive |
| "Gene" | Genotype | Functional range of individual | Individual (phenotype) |
| Genotype | Gene pool | Niche (population system) | Population (deme, cline, species) |
| Gene pool | (Note 1) | Ecosystem (community system) | Community |
| (Note 1) | (Note 2) | Provincial system | Province |
| (Note 2) | (Note 3) | Marine biosphere (world ocean system) | Total marine biota |

Note 1. There seems to be no single term for an organized collection of gene pools.

Note 2. There seems to be no single term for an organized collection of gene pool collections.

Note 3. This is the collection of all the nucleotides in the world.

(Valentine 1973, Table 3-1)

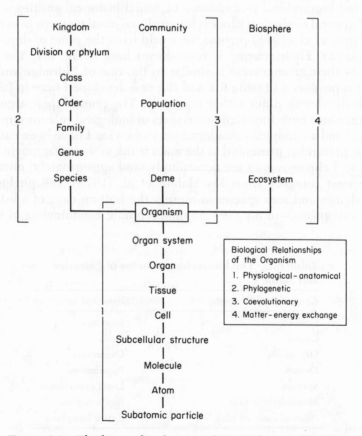

Figure 6.1   The hierarchical views of MacMahon et al. (1978).

may be found in close association on a single bedding plane (surface of deposition). A similar collection, involving representatives of nearly all the same species (with perhaps a few now missing, perhaps some others added), may appear on another bedding plane that is, say, a million years younger or older. Moreover, there is a hierarchical nature to the resemblances between such associations: Paleozoic molluscan faunas of the intertidal zone tend to retain a distinctive flavor from the Ordovician through the Permian, with a typical spectrum of wide-apertured bellerophontid gastropods, nuculoid bivalves, and so forth. The assemblages vary with different substrates, and these variations show definite continuity through time. That there is stability in ecological systems, a kind of temporal persistence that far transcends the sort of stability ecologists have in mind when they invoke the term, is an overwhelming generalization of the fossil record (cf. Boucot 1978), one that we owe most to the work of Valentine (1968, 1969, 1973) and Bretsky (1968, 1969; Bretsky and Lorenz 1969). The hierarchical nature of the resemblance between units through time is a simple function of the hierarchical arrangement of taxa—itself a simple outcome of the evolutionary process.

But the hierarchical resemblance of actual historical entities is not the sort of general ecological process hierarchy of usual concern to ecologists. MacMahon et al. (1978) picture the world from the point of departure of an organism. Their scheme is reproduced here in fig. 6.1. The overall thrust of their arrangement is similar to the one of Eldredge and Salthe (1984) reproduced in table 6.2 and the one developed here in table 6.3, though the details differ rather markedly. The crux of their argument is that organisms have four significant sorts of biological relationships. Physiological and anatomical considerations yield what I have been calling the somatic hierarchy, presented as the main trunk of their diagram in fig. 6.1. Genes and chromosomes are accordingly (and appropriately) missing.

The next category from MacMahon et al. (1978), the phylogenetic, omits demes and sees species as merely the bottom rung of a list of Linnaean categories—in my view an unacceptable commingling of the pro-

Table 6.2   The hierarchical schema of Eldredge and Salthe

| *Genealogical hierarchy* | *Ecological hierarchy* |
|---|---|
| Codons | Enzymes |
| Genes | Cells |
| Organisms | Organisms |
| Demes | Populations |
| Species | Local ecosystems |
| Monophyletic taxa | Biotic regions |
| (Special case: all life) | Entire biosphere |

(Eldredge and Salthe 1984)

Table 6.3    The genealogical hierarchy and two
versions of the ecological hierarchy

| Genealogical | Ecological (I) | Ecological (II) |
|---|---|---|
| Monophyletic taxa[a] | Regional biotas | Biosphere |
| Species | Communities | Ecosystems[a] |
| Demes | Populations | Populations |
| Organisms | Organisms[b] | Organisms[b] |
| Chromosomes | Cells | Cells |
| Genes | Molecules | Molecules |

*Notes:* The difference between the two versions of the ecological hierarchy depends upon the nature of the integration of the abiotic realm. In the center, reflecting the view pursued for the most part in this book, the abiotic realm is seen as a separate hierarchy. Inclusion of the abiotic realm with the ecological hierarchy is figured in the right column.

[a]Individuals themselves hierarchically nested within still larger individuals.

[b]Individuals themselves composed of the somatic hierarchy of individuals, namely cells, tissues, organs, and organ systems.

cess genealogical hierarchy and the historical hierarchy of nested taxa. It is the coevolutionary and matter–energy transformational aspects of organisms that form the thrust of the development of MacMahon et al. (1978), considerations that lead to the two parallel hierarchies on the top right side of fig. 6.1. They see organisms aggregated into demes, themselves nested within populations, which in turn form communities. But organisms are also integrated into ecosystems, which unite to form the biosphere. It is traditional in ecology that the ecosystem explicitly includes reference to the abiotic environment, whereas entities such as communities are collections of organisms alone. Though at first glance it seems that MacMahon et al. (1978) have preserved this distinction in their two parallel ecological hierarchies, in fact they depart radically and dramatically from it. The elements of the glue that imparts cohesion to demes, populations, and communities in their view—the very criterion of membership in their hierarchy—are coevolutionary relationships, interactions of whatever nature that affect the reproductive success of organisms. Thus, their definition of the critical ecological entities is as follows:

*Deme*—a group of coevolutionarily interacting organisms that interbreed in a manner approximating panmixis.

*Population*—a group of coevolutionarily interacting demes or non-outcrossing organisms, which due to a common descent are genetically similar enough to be considered conspecific, yet do not interbreed sufficiently to be a panmictic unit.

*Community*—a group of coevolutionarily interacting populations connected by effects of one population on the demography or genetic constitution of the other.

*Ecosystem*—the plexus composed both of abiotic entities and at least one organism, which are united by the exchange of matter and energy.

*Biosphere*—the ecosystem which includes all of the organisms on Earth during a defined time interval. (MacMahon et al. 1978, p. 704)

Note that of the four entities (including organisms) first considered in the list, the next higher unit consists of interacting entities that are on the next lower level. Indeed, it is the very interactions of the next lower level that defines the successive higher levels in the scheme of MacMahon et al. (1978). Hull (1980) points out, however, that groups are *not* individuals. My more severe objection to their definitions of demes, populations, and communities stems, of course, from their stress on reproductive success. I entirely agree that interactions among populations within communities may, can, and indeed do affect genetic change among organisms within demes. But it seems to me far more useful to see the entities involved with reproduction defined in the more traditional way that I have discussed under the genealogical hierarchy. Thus, the sort of informational interactive effect that MacMahon et al. (1978) see as providing the glue that lends cohesion to demes (in an explicitly ecological context), populations, and communities is better viewed as a cross-hierarchy effect, the main driving force of stasis and change in evolution. To define these ecological units in terms of coevolutionary interactions is in effect to blur the important distinction between information and matter–energy transfer in evolution.

And MacMahon et al. (1978) do acknowledge that the "transfer of energy or mass between an organism and its environment" (citing the words of Spomer 1973) involves another set of interactions of organisms in the natural world. If we leave reproduction (information transfer) to the genealogical hierarchy as a matter of definition, it is precisely such energy–matter interaction that we are left with to define—actually, to impart cohesion to—ecological entities. And this is precisely the approach taken in Eldredge and Salthe (1984), where the basic elements of the ecological hierarchy are seen as molecules, organisms, populations, local ecosystems, and, finally, regional biotas. Accepting, however, the distinction between purely biotic units versus those that commingle the biotic and abiotic (e.g., ecosystems), it would perhaps be as well to revert to the following list of nested entities that form the ecological hierarchy: molecules, organisms, populations, communities, and regional biotas. But such a system raises the need to consider the abiotic world separately, a situation that perhaps makes more sense but complicates the evolutionary picture still further. For considering the abiotic world separately necessitates recognition of still another hierarchy. Yet there is no doubt that the abiotic world can (does!) directly affect individual organisms—a simple fact that MacMahon et al. (1978) recognize when they write (p. 702): "By our definition, ecosystems containing a single organism are conceivable. This is contrary to the traditional view (Odum 1971, Whittaker 1975) that an ecosystem consists of a community plus its environment (which is true in some cases)." However we consider the abiotic world, whether as an integral aspect of some form of ecological hierarchy or as entirely separate from both the ecological and the genealogical hierarchies, it must be seen to impinge upon all levels of the ecological hierarchy.

The abiotic world is complexly structured and itself has hierarchical aspects. I will leave it as a black box. The abiotic world has blatant, obvious effects on both the genealogical (informational) and ecological (energy–matter transfer) interaction of biological entities. After all, the weather can trigger both mating and migration among the same organisms. But it is tempting to see the abiotic world affecting the genealogical hierarchy largely through the ecological hierarchy. This is because nearly all the adaptation that evolutionary biologists have focused on so assiduously ever since Darwin is concerned not with organismic reproduction but with ways and means of effecting energy–matter transfer, of making a living. These adaptations have implications for reproductive success, to be sure: but they are *about* energy–matter transfer. Thus, the abiotic world remains separate in the present scheme, but the ecological hierarchy remains concerned with energy–matter transfer, and the list of entries in the ecological hierarchy includes molecules, organisms, populations, communities, and regional biotas (up to and including a summation of all regional biotas) as the final rung.

But are such ecological entities spatiotemporally bounded individuals? Matter–energy transfer largely takes place on a moment-by-moment basis. We need to ask if such interactions in fact bind these putative entities into units resembling coherent wholes. More-making, the general process that provides the temporal cohesion for next larger genealogical entities, does not enter into consideration of cohesion of the elements of the ecological hierarchy.

Molecules, of course, lie at the base of the somatic hierarchy—the cells, tissues, organs, and organ systems that compose organisms. Indeed, MacMahon et al. (1978) justly add subatomic particles and atoms to the bottom of the list and interpolate subcellular structures as well. There is, in general, no difficulty in agreeing on the discreteness, the particularity, the individuality of such structures—once, at least, the scale barrier is surmounted. And, of course, the intra- and extracellular physiological processes are seen in the light of energy–matter interactions. Indeed, it is virtually impossible to see them in any other way. The hierarchically arranged entities that comprise a multicellular organism are classes of individuals. And each individual is integrated at least in part by the interactions of the next-lower-level individuals; cells have outer limits set by membranes but are further integrated by the interactions among their component organelles. The same holds for all other levels, though the interaction among cells within tissues, tissues within organs, and so on are often most clearly seen in induction sequences during morphogenesis.

Cellular division is an apparent exception to the generalization that individual entities are held together at least in part by the dynamic interactions of the next-lower-level component individuals. Without continual cell division, a multicellular organism dies. There are clearly overtones of reproduction here; indeed, mitotic cell division in purely asexual unicellular organisms (if indeed there are any such—the evidence seems to be against it) *is* reproduction, whereas the process has shifted to "mere" somatic maintenance within sexually reproducing multicellular organisms. Mitotic cell division in multicellular organisms is a ghost of simpler times.

And organisms clearly belong to the ecological hierarchy, as MacMahon et al. (1978) explicitly have stated and virtually everyone else as a matter of course has recognized. Organisms are prima facie individuals. They are just as obviously "economic machines adapted to various ways of making a living . . . in an ecological context, energy exchange drives the system and the individual organism has a standard thermodynamic history (Zotin 1972)" (Eldredge and Salthe 1984). There really can be no argument with the propriety of seeing organisms in this economic context any more than in seeing them as participants in the genealogical hierarchy.

It is once again the higher reaches of the hierarchy that pose the problem for seeing the entities as something more than mere groups. When we go beyond the organism level, we have dropped any semblance of clearcut boundaries analogous to membranes and skins. As in the genealogical hierarchy, it is in these upper reaches of the ecological organization that ontology needs reshaping.

It is commonly said that populations, communities, regional biotas, and the like are difficult to define, when what is meant is that individual, particular instances are difficult to define—in the sense of delimiting or simply recognizing—in nature. Much the same confusion has plagued considerations of species (see above; also Eldredge and Cracraft 1980, chapter 3). There should be some distance between our ontological notions and the purely epistemological problems of observing entities, though, of course, it remains true that the ontology of the relatively large (relative, that is, to the human spatiotemporal scale of existence) is tricky mostly because we lack the methodology to see up as easily as we have been able to see down. (There is no equivalent of the bubble chamber for species or communities; telescopes work on galaxies but are inappropriate for species and communities.) As usual, the higher we go the less real such entities have always seemed to many observers.

Populations, however, are not that difficult. Conspecific organisms tend to flock together for functional reasons beyond reproduction. Indeed, population biology is largely concerned with density, number, growth rate, and so forth—aspects of populations that stem from characteristic interorganismic, within-population behaviors. Pack hunting is but one such interaction with obvious energy transfer overtones. Populations are more than mere collections of conspecific organisms that happen to inhabit the same local area. They are united by more than (actually, by factors other than) reproduction. The existence of populations depends upon additional sorts of interactions among conspecific organisms, such that the population as a whole becomes an economic entity within the next higher level—that of the community.

The boundaries of populations are usually most clearcut when the spatial boundaries of suitable habitats are themselves clearcut. Klipspringer populations on isolated kopjes on the African plains leap to mind. But difficulties in recognizing such boundaries (hence to conceding the individuality of such populations) stem from one further aspect of the ecological organization of organisms: such units themselves are hierarchically arranged. Thus, in some social mammals, an organism may be a member of a family unit (the economic side of a blatantly genealogic entity), several families may ordinarily associate in the daily routine of foraging, and so forth.

In sum, ongoing among-organism reproduction keeps sexual demes going, but nonreproductive organismic interactions are what lend cohesion to populations.

## Communities and Higher-Level Ecological Entities

The ecological hierarchy deals with matter–energy transfer. The genealogical hierarchy, in contrast, is concerned with information, which is ensconced, at base, in the genome. The two are worlds apart, yet thus far the ontological individuals that I have listed in the ecological hierarchy bear a striking resemblance to lower-level members of the genealogical hierarchy. From at least as far back as Weismann, we have distinguished cells of the soma from those of the germ line, but at the organism level there is a perfect isomorphism, really an identity: each particular organism is simultaneously a member of both the ecological hierarchy and the genealogical hierarchy. At the level of populations (ecological hierarchy) and demes (genealogical hierarchy), the individual entities again almost lose their separate identities; the common inclusion of demes in ecological discourse (cf. MacMahon et al. 1978) underscores the fact that demes and populations are near enough in composition to be confused, or at least for demes to be regarded as parts of the populations. Yet all these entities in the two hierarchies are arguably near enough that all we could be talking about so far is different functions or aspects of the same or similar entities—matter–energy interactions in an ecological sense, information transfer in a genealogical sense.

It is at the blatantly cross-genealogical level of communities that the independence of these two process hierarchies becomes utterly clear and justifies the recognition of two separate hierarchies in the lower rungs of each, where the distinction between the nature of the individuals in each is perhaps less clear. Communities are associations of populations of nonconspecific organisms. Even if the definition of MacMahon et al. (1978) were adopted (where communities are defined and recognized by coevolutionary interactions among populations), the nature of these coevolutionary interactions—involving competition, predator–prey relationships, and the like—obviously include matters of energy–matter transfer. Restriction of consideration of the exchange of matter and energy only to an independent hierarchy of ecosystems and the biosphere

(MacMahon et al. 1978) that includes the abiotic realm begs the question of what is going on in the coevolutionary interactions of the parallel ecological hierarchy that MacMahon et al. recognize. If two nonconspecific populations are competing, and thereby affecting the reproductive success of component organisms within each, they are competing not for mates but for such items as food and shelter.

A related and perennial problem in the very definition of communities is the amount or degree of such interaction requisite in a community. Some ecologists have preferred to define communities almost strictly as recurring associations of nonconspecifics, while others have stressed the interactive interconnectedness of such associations. Clearly there is a spectrum, though the extreme possibility that a similar configuration of nonconspecific organisms is repeatedly assembled given only (1) the ability to reach (colonize) a given area and (2) a repetition of a particular physicochemical (specifically abiotic) regime seems unlikely. All animals eat some other organisms. I presume there is some level of complexity of among-population interaction within every community.

But the notion that communities are recurrent associations is arresting. For it suggests that in spite of the among-population interactions that serve to bind a community together, if a particular community is destroyed, something very like it may once again be assembled—another community so similar to the first that a paleoecologist would be forced to consider it the "same" were both to be found in the fossil record.

In the genealogical hierarchy, the relatively larger entities—species, monophyletic taxa—are spatially more broadly distributed and characteristically persist longer than the smaller subunits. The same appears to be true of communities vis-à-vis lower ecological entities such as populations and organisms. Populations may disappear, altering the web of life, yet, at least in the view of communities adopted here, the community itself persists. Rather like species, which never remain wholly static with respect to the genomes and collective phenotypes of component organisms (and demes), communities can be imagined to persist as entities without the expectation that their components never change. Thus, communities appear to have more than a momentary existence.

And yet there appears to be recurrent association, very like modern communities, in the fossil record. And it cannot be known with assurance that when such an assemblage of populations, identified as a community, reappears in rocks two million years younger than those that house the nearly identical assemblage, the association persisted throughout the span, tracking the migrating substrate (habitat) that is indeed the predominant mode of environmental change through geological time. The species persists, but whether or not the community itself persists is another matter. Presumably, sometimes it does and at others it is truly a reassembly of actors given the return of suitable environmental conditions and the continued existence of the species that enable such reassembly to occur. In this largely speculative discussion, and in this context and this sense only, communities, if they cease to exist yet seem to reoccur, are classlike entities.

The difficulty in seeing persistence of ecological entities in geological time—because of the distinct possibility that we are in fact taking classes for persistent entities—underscores the main point of evolutionary relevance of all ecological entities: it is not their temporal stability but their spatial distribution and, above all, the momentary nature of their internal interactions that are of greatest evolutionary significance. Thus, it is the constant interactions among populations that creates a community; it is the continuity of those interactions that provides whatever temporal stability such entities may in fact display.

Earlier I alluded to a hierarchy of resemblance among ecological associations through geological time. The obvious hierarchical nature—nested arrangements of co-occurring, nonconspecific organisms—is one of several historical hierarchies that form simply as the outcome, the direct fallout, of the evolutionary process. The relevant point here concerns an earlier observation of the nature of species: species are not ecological entities, not being confined (certainly as the overwhelming generalization) to any single ecological entity (save larger-scale regional biotas, and certainly the entire biosphere.) Moreover, as the complexion of communities clearly shows, the various species that form the nonconspecific populations that form a community also typically contribute populations to different communities elsewhere. White-tailed deer (*Odocoileus virginianus*) occur in a rather wide range of habitat, while the various mice, gallinaceous birds, and so forth whose local populations ("avatars," in the sense of Damuth, in press) form the local interactive system I am here calling communities (realizing that some would prefer "local ecosystems") of which white-tailed deer may be a part, are not everywhere found in association with *O. virginianus*. Coyotes (*Canis latrans*), on the other hand, co-occur and interact with both mule deer (*O. hemionus*) and whitetails, as well as a host of other species. There is regularity in patterns of co-occurrence of nonconspecific populations, enough for us to recognize spatiotemporally localized entities such as communities. But there is also a kaleidoscopic pattern of interchangeability of species membership that conveys a hierarchical pattern of similarity linking communities (or local ecosystems) into larger interactive systems. Thus, local systems are bonded into regional systems and so on, until we reach the level of the worldwide biota. Just as monophyletic taxa together form the upper-level class (i.e., kind of individual) of the genealogical hierarchy, the cross-genealogical combination of interacting nonconspecific populations of organisms forms the general class of cross-genealogical entities of the ecological hierarchy. Yet, just as some monophyletic taxa are of a larger scale and indeed subsume others, there are ecological entities on a larger scale that subsume others. And the interactions among local communities that integrate more regional systems may well be of a nature rather different from the among-population interactions that integrate a local community.

Eldredge and Salthe (1984) have discussed the difficult conceptual problem of visualizing the interpopulational interactions in nature. They write:

Populations—not organisms—interact at the ecosystem level [i.e., what I am here more simply calling the community level]. This is tricky: despite the fact that one might observe a fox killing and eating a quail, it is nonetheless populations that are interacting. This is a matter of theory, and we must recall the principle of non-transitivity. The singular act of a given individual with respect to another of a different species, so vivid to our senses, has no more theoretical import at this level than have the mutual surges of hormones and emotions accompanying mating within two organisms (or the jostling of the liver of the one by the motions of the other) for the population level. When we observe this act of predation we are seeing it through the filter of a theory that is not appropriate in our present context. Then, too, there is the matter of scale. In order to report correctly our observation of this event from the present perspective we must take the trouble to note that a part of the predator population is culling a part of the prey population, if, indeed, we choose to report the event at all. But in order to do this with facility we must have a mental paradigm switch. Without that it will be easier for us spontaneously to observe events on our own scale which may be correlated with what is going on at, say, the ecosystem level than to see these latter whole (i.e., holistically)—something that, alas, may not be really possible. Part of the problem of scale concerns time. The relaxation times and equilibrium constants of processes at a higher level like this are considerably longer and larger than those applicable to the organismic events we are familiar with. (Eldredge and Salthe 1984, p. 197–198).

What happens to the individual fox and the individual quail is of material relevance (possibly) for the content of the gene pools of both the local fox and the local quail demes. What matters for the community, though, is simply that there are foxes eating quails.

## THE ECOLOGICAL AND GENEALOGICAL HIERARCHIES

The ontology of the larger sorts of biological entities—species, communities, monophyletic taxa, and the like—has provided most of the difficulties in evolutionary theory. Perceptual problems arising from scale alone account for much of the difficulty we have had in understanding the nature of such units and seeing their role in the evolutionary process.

The positive side of these relatively intractable entities is that, however dimly we may perceive their nature, merely acknowledging their presence at all strongly underlines the importance of such general themes as matter–energy transfer and information—or put in perhaps less stiff if equally jargonish ways, Hull's interactors and replicators.[3] It may be difficult to tease these two fundamental aspects of biological systems apart—except purely conceptually—in such lower-level entities as cells and organisms, but these same rather unwieldy larger items—these demes, populations, species, communities, and so forth—are *organized* around this fundamental dichotomy in the organization and functioning of biological systems. However inadequate my particular attempt to describe the general properties of these systems may be, there can be little doubt (as so many authors before me have struggled so mightily to establish) that life is in fact organized *simultaneously* in these two very different ways.

Evolutionary theory simply has not taken this organization of life fully into account in its attempt to explain what evolves and how it evolves. Indeed, the very focus of evolutionary theory ever since Darwin has been the relationship between economic adaptations and relative reproductive success—legitimate in its own right but contributing to our collective failure otherwise to distinguish these two functions (economic and reproductive). Once again, the issue is ontological. Ghiselin's (1974b) *The Economy of Nature and the Evolution of Sex* specifies the two functions in its very title but fails to distinguish between them in its narrative: according to Ghiselin, economic competition is competition for sex.

The modern synthesis focuses, roughly, on one-half to two-thirds of what I have been calling the genealogical hierarchy. Studies in coevolution (see Futuyma and Slatkin 1983 for a recent compilation), where the interactions among nonconspecific populations are seen explicitly as affecting the genetic composition of demes, more directly link genetic with ecologic systems. This is a healthy step. But a complete evolutionary theory must recognize *all* the entities of *both* hierarchies. There are processes going on within both systems that are quite independent of processes in the others. It is these processes plus the cross-interaction between elements of the two process hierarchies that produce the historical pattern of events—in addition to changes preserved in organisms themselves—that led to the very notion of evolution in the first place. I shall now consider the general nature of these within- and especially among-hierarchy processes.

## NOTES

1. It is fashionable nowadays (see Sober and Lewontin 1982 for but one example) to warn against the perils of "reification" and "entification," that is, falling prey to the error of seeing some phenomenon or other as real where in fact it is only apparent. We have here the old distinction between type I and type II error in statistics. The potential error of reification is saying something is there when it "really" isn't. I am concerned with the converse: treating real things as if they don't exist—to my mind the more serious source of error in the actual practice of contemporary evolutionary biology.

2. All placental mammals so far known have hair, mammary glands and three middle ear bones. This does not mean that all such features were "invented" at one and the same time, all within a single ancestral species. Indeed, evolution usually adds novelties seriatim, resulting in what is usually called mosaic evolution.

3. Thermodynamics and evolutionary theory have had a long, if as yet generally unfruitful, mutual intellectual history. Clearly the systems—within levels within each of the two major hierarchies, between levels within each of the hierarchies, and between the two hierarchies in general—lend themselves to description by the equations of thermodynamics or information theory. Wiley and Brooks (1982) provide a recent entry into the connection between thermodynamics and evolution.

# 7

# Hierarchic Interactions:
# The Evolutionary Process

"It is the intertwined and interacting mechanisms of evolution and ecology, each of which is at the same time a product and a process, that are responsible for life as we see it, and as it has been." (Valentine 1973, p. 58)

Organisms—biology begins with organisms, and indeed a great deal of the history of biology is a trek through progressively finer subdivisions of organisms. When "forefronts" of biology are listed these days, nearly all concern the molecular biology of intracellular (and intraorganelle) physicochemical processes—and quite rightly so. But the ontology of units larger than organisms, while not wholly neglected, is at least as difficult a problem. Organisms are by far the easiest of biological units for us to see, to probe, to conceptualize as "individuals."

But, in the present context, organisms pose a unique problem all their own: they constitute the only class of individuals to be found in both the genealogical and ecological hierarchies. Consider the confusion that permeates even the recent explicitly hierarchical literature: ecology and evolution (as in the quote from Valentine that stands at this chapter's head) are generally seen as separate areas of inquiry, but the choice of the higher-level individuals to be incorporated into one's hierarchy very much depends upon one's point of view.[1] Below the organism level, of course, the distinction between the somatic and germ lines (i.e., in multicellular organisms) once again ensures a clean separation of the elements of the two hierarchies. Hence the conclusion (Eldredge and Salthe 1984) that there must in fact be two independent, yet parallel and interacting, process hierarchies that together combine to yield evolution.

Organisms, as members of both hierarchies, threaten to muddy the picture. It is possible, of course, to distinguish between the economic and reproductive functions of organisms, as I have done at length in the preceding chapter. Physiologists, after all, have long been telling their students that reproduction is the one physiological process not essential to the survival of an organism; thus, it is no surprise that it is invariably the first such process to be dispensed with when the organism is stressed. It is easy to distinguish the economic from the reproductive functions of the vast majority of organisms, but in many vertebrates, most especially *Homo sapiens*, sexuality has clear economic implications, obscuring the distinction between the two hierarchies perhaps even more. The connection is closest where sexuality (as in demes, not particularly in social insect colonies) is directly associated with sociality (i.e., contributes to the cohesiveness of the social unit). This, of course, raises the issues of kin selection (Hamilton 1964a, 1964b) and sociobiology (Wilson 1975). For the vast majority of organisms, such commingling of reproductive and economic functions simply does not occur.

I belabor the distinction between the genealogical and economic hierarchies here because it is the goal of this chapter to integrate them—or, rather, to investigate the interrelationships between them. The fact that organisms belong to both hierarchies then becomes interesting, even vital. But there is nothing even remotely approaching identity between the two hierarchies as systems. The picture, instead, is of two different hierarchical systems mutually dependent for their very existence. And by existence I mean both the *persistence* and the *modification* of every individual of each class in both of these hierarchies. The resultant products of such interactions—evolution—is just as Valentine says: "life as we see it, and as it has been." The ecological and genealogical histories of life can be properly viewed as strongly linear, or as various hierarchically arranged patterns—such as the taxic (Linnaean) and homology hierarchies, plus the unnamed historical hierarchy of ecological systems—which we may, with some justification, label the Bretskyan hierarchy.

## PROCESSES WITHIN THE GENEALOGICAL AND ECOLOGICAL HIERARCHIES

I have already identified the two classes of process that distinguish the genealogical from the ecological hierarchies. In the genealogical hierarchy, the very criteria for inclusion, and the very processes that lend cohesion to the individuals of the next higher level, are birth and death, the ongoing more-making of individuals "of like kind." In the ecological hierarchy, in contrast, we are concerned with the interactions between individuals of like kind (such as organisms within populations, or populations within communities)—interactions that commonly, if not invariably, are connected to energy conversion or transfer. It is such interactions that lend cohesion to the next-higher-level individuals within the ecological hierarchy.

Biases in births and deaths of individuals within the genealogical hierarchy produce change within that hierarchy, while within the ecological hierarchy change depends ultimately on the identity of the interacting individuals. The source of the bias of births and deaths within the genealogical hierarchy is commonly the ecological hierarchy (i.e., commonly but not invariably—cf. Dover 1982 on molecular drive, or even Wright's genetic drift). And the source of change in the composition of the inter-active players in the ecological hierarchy is, of course, the genealogical hierarchy. The abiotic world, so strongly implicated in both stasis and change in both hierarchical systems, is seen here as acting most directly on the ecological hierarchy—indeed, forming a part of that system if we choose to include ecosystems rather than communities in that hierarchy.

But before I characterize the nature of these simultaneous and complex interactions between the two hierarchies, the processes within each—particularly the interlevel processes—require additional clarification.

Vrba and Eldredge (1984) have contrasted birth and death processes among levels within the genealogical hierarchy. Replication of genes, duplication of chromosomes, organismic reproduction (asexual and sexual), and speciation (plus the more-making of demes) are all vastly different biological processes, though their commonality as ways to produce "more individuals of like kind" is clear enough. For example, organisms possess adaptations for reproduction, while species do not. Speciation represents a disruption of organismic reproductive adaptation (Vrba and Eldredge 1984). Similarly, death processes are different among the various levels of the genealogical hierarchy: organisms possess internal causes for death, while species do not. Indeed, though Eldredge and Salthe (1984) present a quasi-hypothetical example of a species' extinction event (as well as that of all its included demes) that does not entail the deaths of all organisms within that species (an event occasioned by the loss of any further chance for the exchange of gametes), it would seem that the deaths of all individuals higher than the organism level within the genealogical hierarchy are realistically reducible in a fairly straightforward manner to the deaths of its organisms. But this may be mere common-sense ontology intruding once again.

The cause of the deaths of individuals up and down the genealogical hierarchy leads to the general problem of what Campbell (1974) calls upward and downward causation in hierarchical systems. Indeed, a major source of asymmetry within this hierarchy—at least potentially—comes from the consideration of death. Clearly, death of an individual at any level automatically implies the death of all its included parts, all its component, lower-level individuals. Externally induced death of an organism, say, a quail in a fox–quail interpopulation predator–prey interaction, kills off that quail's lower-level components as well. Extinction of a species, it may readily be imagined, must entail the deaths of all its component demes, organisms, chromosomes, and genes. The problem, of course, arises when we try to imagine how that might happen; as I argue in chapter 6, species are not interactors and hence not members of the ecological hierarchy.[2] Species are, instead, repositories of information, cohesive reproductive communities. They do have births (speciation) and deaths (extinction), but the latter can only result from the cumulative deaths of all component demes, a function of the deaths of all reproductively active organisms belonging to those demes.[3]

But the asymmetry principle still remains: as high up in the genealogical hierarchy as one can admit the possibility of directly externally induced death, all subcomponents will automatically die, too. The converse is not true: in the genealogical hierarchy, the existence of $L_i$ individuals depends upon the ongoing production of new $L_{i-1}$individuals and not upon their precise identity. Thus, $L_{i-1}$individuals are born and die, changing the characteristics of the $L_i$ individuals, but in a far less profound manner than the downward causation of the deaths of higher-level individuals. *Sorting of lower-level individuals, in general, has a far less profound effect on upper-level individuals than the converse.* The discovery in the 1960s that allelic heterogeneity is even greater than was previously suspected has been augmented by the observations of molecular biologists (e.g., Dover 1982) that parts, at least, of the genome seem to be in a nearly perpetual state of flux—without, for the most part, greatly affecting the organismic phenotype or, consequently, any still higher level in the genealogical hierarchy. Yet, forgetting for a moment how they became extinct, the demise of North American elephants only a few thousand years ago greatly reduced the amount of extant genetic variation in the elephant clade.

Levins (1968) lists three desiderata of models: generality, realism, and precision. He makes a convincing case that in any one instance we can expect to maximize but two of these three factors. Here, I think, realism must remain uppermost and is best served if we do not become overly formal in our hierarchical expectations of nature. I have in mind the dichotomy broached in chapter 1: the alternatives of reductionism and hierarchy as an approach to nature's complexity and as a structural hat rack on which we frame our explanatory process theories of evolution. In examining the synthesis, I found that in earlier stages it was not as single-mindedly reductionist as it later became, or as it has been made out to be. Indeed, we can point to Dobzhansky's (1937a) development of hierarchy as explicitly antireductionist, at least to a degree. In eschewing such perceived reductionism, though, and adopting in its stead a hierarchical approach, it is tempting to stress the independence, or quasi-independence, of processes among levels within each of the two hierarchies. This would entail the strict application of the principle of nontransitivity (see Eldredge and Salthe 1984; Salthe 1985) and perhaps would amount to as great a distortion of the "true" organization and workings of nature as an obligate reductionist approach. Some processes within the genealogical hierarchy are starkly different and independent at the various levels, most notably those involving the births of the various individuals. Deaths of the very same individuals, however, realistically do seem to be transitive to a great extent. Once again, we must be careful to segregate cause and effect. North American mastodons and mammoths became extinct because their numbers—that is, of component organisms—dwindled below the critical mass needed to maintain demes within each of the species involved. The effect of this was greatly to impoverish the genetic diversity of the world-wide clade Proboscidea and to bias the representation of genes within Placentalia, Mammalia, Tetrapoda, Deuterostomia, Metazoa, Animalia, Eukaryota—and thus within all of life.

Much the same might be said for interlevel relationships within the ecological hierarchy. The interactions between component individuals that lend cohesion to individuals of the next higher level are intrinsic to each particular level; the interactions among organisms within a population are utterly different from those that go on among populations. Yet (and cf. discussions in chapter 5 on nontransitivity) it is obviously organisms that are performing the interactions when we observe events of interpopulation interaction. The complexity of the situation emerges when we examine niche theory. Though it is sometimes said that species have niches (they do not), the ecological literature generally discusses niches at the population level. Yet niches patently involve adaptations, and these, of course, are phenotypic (morphological, physiological, and behavioral) attributes of organisms; features, moreover, clearly under (morpho-) genetic control. Enter the genealogical hierarchy immediately, in what was (niche theory) purported to be an ecological concept at the population level. Though niche theory remains at base a population concept, it also involves the properties of individual organisms plus aspects of communities, as well as the abiotic realm (or, if you will, ecosystems). Niches cannot be construed strictly with reference to one single level of the ecological hierarchy; rather, populations are the focal level, and the next adjacent (lower and higher) levels are directly involved (as a triad—Eldredge and Salthe 1984; Salthe 1985) in our very approach to understanding what niches might be. Indeed, adaptations as ecological devices at the organismic level also involve more than one level of the somatic hierarchy. But visualize an organism with the requisite enzymes to digest a particular food substance. It also has the behavioral and anatomical features requisite to find and ingest said food item. It lives where the food item lives, hence belongs to an appropriate community. Food might be there, but some of the lower levels (including, at base, the enzymes) might not be, or the enzymes might be there but any of the various higher-level features are absent. All must be there for this particular kind of economic interaction to work. All cogs in the economic hierarchical wheel must be in place for any particular aspect to function. Patently, processes at any one level within the ecological hierarchy have direct implications for processes at lower and higher levels. It is not a matter of reduction to the lowest level or subsuming under the highest level. A break anywhere in the chain of interconnected levels renders the interdependent activities of the other levels superfluous. Thus, the quasi-independence of processes within the ecological hierarchy: the processes at each of the different levels by nature are in themselves radically different, yet for any one process at any one level to take place typically requires conditions at levels both above and below to be met—the constraints and boundary conditions so commonly encountered in formal hierarchy theory (Eldredge and Salthe 1984; Salthe 1985).

## INTERACTIONS BETWEEN THE ECOLOGICAL AND GENEALOGICAL HIERACHIES

If we must agree with Dobzhansky (1973) that "nothing in biology makes sense except in the light of evolution," it is also true that we cannot embrace a truly general concept of evolution without including both great classes of process: the ecological and the genealogical. We need an understanding of the integration of the interactive processes of energy exchange and transformation and the genealogical processes of information conservation, change, and transmission. If we are accustomed to defining evolution as permanent change in genetic information, we must also realize that both retention and modification are parts of the evolutionary process, and most of the regulation of such retention and change comes from outside the genealogical system—it comes, in fact, from what I have been calling the ecological hierarchy.

The general mode of interaction, the general scheme of interrelationships between the two hierarchies, is simple enough. Looking at the world at any one place and moment, we (as organisms) see other organisms, members of local populations, living in some integrated fashion in communities that typically (invariably) contain other nonconspecific organisms. Such an ecological individual at the community level, from a genealogical point of view, is a collection (a mere set or class) of organisms drawn from various source species. Monophyletic taxa supply species, which in turn supply the organisms, that we observe in each community, while the communities are themselves spatially integrated into the larger units of the ecological hierarchy. *The genealogical hierarchy simply supplies the players in the ecological arena.*

The nineteenth-century use of the term *ecological vicar* illustrates the point vividly. Wherever one species replaces another geographically and the niche characteristics of the respective member populations seem similar, we have in effect a putative case of ecological substitution. One species seems to be the ecological equivalent of the other—ecological equivalence meaning, of course, "the adaptive phenotypic properties of their component organisms." Through geological time the effect is often even more pronounced. Eldredge and Salthe (1984) use the example of bellerophontid gastropods—Paleozoic and Lower Mesozoic archeogastropods. The Devonian *Retispira leda* and related synchronic species are, in effect, replaced by successive suites of congeneric species throughout the Upper Paleozoic, seemingly in equivalent ecological settings (i.e., communities). The effect is truly one of substitution of comparable organisms rather than any stately progressive evolution, even though the time span of the genus *Retispira*—some one hundred twenty million years—is enormous. The point is that for all those years there was yet another species of *Retispira* to fill the ecological bill. Realistically, of course, it is ongoing speciation that kept the genus *Retispira* going, not vice versa, as a careless construal of the present point might seem to suggest. Yet it is true that once speciation within the *Retispira* clade ceased, the possibility that a population of *Retispira* organisms would ever again be found in a muddy-bottom intertidal community was forever gone.

And the effect is magnified the larger the genealogical unit being considered. Final extinction (in the Triassic) of the last bellerophontid species forever removed bilaterally symmetrical herbivorous snails—snails that possessed that peculiar concatenation of plesiomorphies and apomorphies that stamped and defined the bellerophontids among gastropods—from the world's marine biota. Not only were the last vestiges of a significant segment of gastropod genetic variance finally gone, but the complexion of all ecological systems of which bellerophontid populations had ever formed a functional part, from reefs to continental slope, was forever changed. Ecological systems above the level of organisms have their own self-organizing processes—the interactions of various sorts among organisms, among populations, among communities, and so forth. But ecological systems must take what "central casting" sends them, there to pick and choose what will fit in and what will not—as in the often dramatic turnover in species composition (membership) often graphically shown in the successional stages of a sere.

The constraints on which organisms occur where are in general well known and need no exhaustive discussion here. Suffice it to say, though, for the record, that the reason why there are no bears in the forest ecosystems of east central Africa is that for some set of reasons rooted in the physical (abiotic) world, the Ursidae never got there or, at best, were only rare members of any African ecosystem. It is not ipso facto because no particular form of bear phenotype could be imagined to fit into such an ecosystem; in fact, bears do well in what appear to be similar ecological settings in Asia and South America. Ecologists, in general, are aware that the reasons why they see the particular organisms they document living in any particular situation may, and often do, transcend the purely mechanistic notions of ecological integration. A lumbering, sometime-herbivorous carnivore might do very well in the cool, forested slopes of Uganda and Rwanda.

The contribution of organisms from various species (and monophyletic taxa) to ecosystems is not, of course, a one-shot affair; it is constant. It appears trivial that some members of a population of conspecific organisms living within a particular community take the trouble (expend the energy) to reproduce; hence ecologists see no difficulty in equating demes with populations or, at least, of seeing the one as a subset of the other (cf. MacMahon et al. 1978). But consider such community systems over a more protracted temporal span. On the shorter end of the scale, Johnson (1972) documented the migration of a sandbar over a well-established (and studied) benthic community in Tomales Bay, California. The sandbar took a few years to pass through; as it migrated, the community was destroyed and a new one reestablished in its wake. The community was downgraded in the sense that it was reestablished at a lower order on the normal successional scale for such ecosystems in that region. The point here is that those newly reestablished organisms—all of them participants in the entire resuccession culminating in the reforming of the Tomales Bay benthic community (in the area downgraded by the sandbar)—had to come from somewhere. Larval recruitment (probably, in this case, from no very great distance) supplied the new organisms, and supplies us as well with a good, low-level case of how species act as reservoirs, genetic banks that can be drawn upon for the establishment of populations in local areas.

The effect is similar to Dobzhansky's (1937a) point concerning the value of a reservoir of genetic variation within species as a hedge against extinction. Even though we note that species are not ecological entities— direct interactors in the economic sphere of life—it is appropriate to see a species as a collectivity of the adaptive properties of all its included organisms and thus as a source of material for the rebuilding of any severely downgraded (or simply eliminated) local population—a true preventive against extinction. Indeed, extinction of a species involves the deaths of all of its organisms, which means all of its populations living in all of its different ecological settings, however monomorphic or varied these may be. That is why it is so much harder to eliminate a deme than an organism, a species than a deme, and a monophyletic taxon than a species. Faced with extreme environmental distress, the reservoir of genetic variability that Dobzhansky (1937a) saw as a useful contingency could conceivably play a role, but the sorts of patterns of recruitment following such local ecosystem collapse illustrate more the simple reservoir view of species as the more direct and effective antidote to impending extinction. Hence the role of the genealogical vis-à-vis the ecological hierarchy: the genealogical hierarchy is the ongoing supplier of the organisms that form the various ecological individuals. In times of ecological status quo, this amounts to little more than the ongoing production, in situ, of new organisms, and the distinction between the two hierarchies is blurred, or at least conveniently ignored. But following the mass extinctions so commonly seen in the history of life, recruitment for the reconstruction of ecological systems becomes an issue of major importance.

If the genealogical hierarchy supplies the very players in the ecological game, it is that very ecological game that determines, in large measure, what exists in the genealogical hierarchy, which of the particular individuals at the various levels can survive, and in what form. Stasis and change in the genealogical hierarchy are regulated to a very great degree by the processes and events in the real-world arena of the ecological hierarchy. Such an intimate feedback system is, of course, complex and complexly confusing. My position here is that neither hierarchy is prior to the other. Also, given sexually reproducing entities (organisms), *all levels of individuals above the organism level will automatically exist as stable entities in the genealogical hierarchy.* But the continued existence and complexion of higher-level ecological entities depend upon what is available in the genealogical hierarchy, just as the nature of those units in the genealogical hierarchy depends very much on past conditions within the ecological hierarchy.

Paraphrasing Huxley (1958), the elements, the various levels of individuals, of the genealogical hierarchy, once born, can either persist in unchanged form, persist in some modified form, or cease to exist altogether. Their births, their persistence (whether in modified or unmodified form), and their deaths all depend to a very great degree on those events and processes of the ecological hierarchy. And here, once again, it matters very much how we conceive of the ecological hierarchy. I visualize it here, for convenience, as a combined biotic and abiotic world. Were we to visualize the abiotic hierarchy as a separate system, it would be feasible to examine the origin of adaptations to purely mechanical aspects of the abiotic environment—swimming, flight, oxygen extraction and within-body transport, and so on—strictly within some unit of the genealogical hierarchy, such as species or, more accurately, demes. After all, such adaptations for the most part must arise through interaction between a population–deme and the abiotic environment. But it is more generally useful, I think, to view the acquisition of adaptations—phenotypic (behavioral, physiological, and anatomical) innovations—in the context of niche exploitation. In so doing, of course, we are talking about (conspecific) populations within the ecological (economic) hierarchy. There is a further caveat: as I have been stressing, there are two classes of adaptations—*economic* adaptations (concerned with niche exploitation) and *reproductive* adaptations (those that pertain strictly to more-making within the genealogical hierarchy). Both categories of adaptation are most easily observed as properties of organisms, though they may not be wholly restricted to individuals of that level (in both hierarchies).

A very clear example of the effects of the ecological hierarchy on elements of the genealogical hierarchy, of course, is natural selection. The nexus between the systems comes from the fact that some of the interacting conspecifics within a population form a deme, and, as Hull (1980) so clearly analyzes, the unequal results of the interactions show up in the differential reproductive success of those organisms within the deme. The phenotypic attributes of the more successful interactors are represented in their greater fitness, hence greater genetic representation in the next generation. This (i.e., natural selection) continues to be the only deterministic theory of the origin, maintenance, and modification of adaptations at the organism level, unless one counts the spontaneous creation of phenotypic features in toto (e.g., by sudden and proportionately large-scale genetic change) that is presented to and either retained or rejected by the environment. For present purposes, these two possibilities (i.e., natural selection working on generally small and generally intergradational phenotypic differences among organisms within a population–deme versus sudden genetically induced change) are end members of a spectrum of interactions between environment (the ecological hierarchy) and the phenotypes of organisms.[4] I have already given two basic reasons why, I feel, selection operates only at this (among-organism, within-population) level: (1) the difficulty of showing deme- and species-level adaptations (wholly impossible, I think, for economic adaptations; arguably possible, but dubious, for reproductive adaptations) and (2) the simple fact that most species are found to be contributing constituents to more than one ecological individual at any particular level (most notably, populations to more than one community) and therefore cannot be construed themselves as being interacting ecological units.

So much for the modification of gene frequencies, which is what most of us mean by the word *evolution* and patently the phenomenon that the synthesis is primarily geared to address. In handling the origin, maintenance, and modification of adaptation in a few sentences, I do not mean to minimize the phenomenon. But so much attention has indeed been paid to it that adaptation is, at least conceptually, fairly well understood. But perhaps the major problem within the synthesis has been that adaptation was seen as the central theme of the entire evolutionary process. Commingled with a slightly garbled ontology of the larger-scale biotic elements that it includes and hampered by the absence of those elements it ignores, the theory offers an intrinsically distorted picture of the biologic realm and the nature of the evolutionary process. Seeing adaptation through natural selection as the main pattern–process of evolution (as Dobzhansky said was true of Fisher, and which certainly became true of the later stages of the synthesis itself) in particular has obfuscated the real natures and evolutionary roles of (particularly, if not exclusively) these higher-level entities. It has always been these extrapolationist portions of the synthesis—for example, seeing species and by extension monophyletic taxa as simple products of within-population adaptational change— that have proven hardest to accept (Eldredge and Cracraft 1980; Gould 1980a, 1982b). Besides change, we have nonchange plus the births and deaths of all those genealogical and ecological individuals to consider in evolutionary theory.

I have said that the births and deaths of ecological units depend very much on events in the genealogical hierarchy, on what is available to "staff" the ecological units. This is true mostly for (ecological) births, for "recruitments." It is possible that infectious disease, for example, can cut across ecological units, striking all organisms within a single species and thus causing its demise—unlikely but possible. If that species were a contributor to a vital part of the trophic structure in one or more ecological units, the extinction of the species could be said to have triggered the demise of the ecological unit. The reverse, however, is more commonly the case: perhaps the greatest signal in the linear history of life comes from the numerous cross-genealogical extinction events, which vary greatly in their severity as measured in several ways: areal effect, number of habitats involved, and numbers of demes, species, and monophyletic taxa included. Some extinctions are local and minor, such as Johnson's

(1972) description of the sandbar in Tomales Bay. Others are more far-reaching: not only elephants (several species) but rhinos, sabertooths, and so forth, no longer roam North America as they did a scant eight to ten thousand years ago. Indeed, some of them no longer roam the earth. And some of the very big extinction events nearly wiped out all of life (Newell 1963; Raup and Sepkoski 1982), such as the Great Dying at the end of the Permian, which Raup (1982) estimates involved the extinction of perhaps as many as ninety-six percent of species then existing. Clearly, such events are of enormous evolutionary significance—talk about the loss of genetic variance!

Here the actual (Linnaean) ranking of monophyletic taxa becomes important, as it was in the bellerophontid example above and for the very same reasons. Taxa ranked relatively high—orders, classes, phyla, and so forth—simply tend, at any given moment, to have more species than do their included taxa.[5] Thus, the more highly ranked taxa tend to be more immune to extinction. If we did lose more than ninety percent of the world's species in the terminal Permian extinction, according to Sepkoski (1980), among well-skeletonized marine invertebrates we lost progressively fewer families, orders, classes—and no phyla.

Ecosystem collapse—of whatever order of magnitude and frequency—does not appear to spring from events within the genealogical hierarchy. Whether the putative mechanism for any particular instance is purely abiotic (such as the Alvarez et al. 1980 hypothesis of meteoritic impact as the trigger for the terminal Mesozoic event) or involves a more complex collapse of trophic structure, the causes nearly always appear to arise within the (broadly conceived) ecological hierarchy. *The deaths of genealogical entities above the organism level are generally caused by events in the ecological hierarchy.* Deaths of organisms occur by both extrinsic (ecological) and intrinsic (e.g., senescence) agents. In any case, the soma is best construed as an ecological (economic) entity; sex cells die spontaneously or through the higher-level death of the organism.

Births, too, of genealogical elements above the organism level are largely to be construed as a reaction to events and processes within the ecological hierarchy. Thus, the origin of demes is very much an ad hoc dependence upon the establishment of new viable populations of conspecifics. Moreover, speciation, even in the conventional synthesis, has come to be seen as an accident of changing physical and biological (largely the former)—hence environmental—parameters. Births of new monophyletic taxa are marked by the acquisition (in organisms belonging to the ancestral species) of one or more phenotypic novelties or modifications of preexisting novelties. These become the synapomorphies shared only by organisms within the ancestral species and by all its descendants. Even if one were to follow a more strictly adaptationist metaphor for the origin of monophyletic taxa, where genetic and somatic architectural constraints and possibilities are seen to combine to yield some key innovation (as has been argued for the artiodactyl astragalus, for example; Schaeffer 1948;

Simpson 1959a) or general body plan (close to the concept of bauplans, e.g., the pentameral symmetry of all but the most primitive echinoderms—Haugh and Bell 1980) that allows or triggers a radiation, the model still involves an adaptive response to some set of external, environmental conditions. Birth (more-making) and death are processes intrinsic to the various levels of the genealogical hierarchy (as detailed in chapter 6). But each such event, at least insofar as individuals above the organism level are concerned, seems in large measure to be a reaction to events taking place in the ecological hierarchy.

Lastly, there is maintenance of the status quo within individuals of the genealogical hierarchy. Stability can be measured in many ways. For example, Eldredge and Cracraft (1980) view the comings and goings of species within a clade (e.g., Miocene–Recent placental mammals, with hominids and a few other taxa among the few exceptions) as constituting a form of status quo. Species diversity may remain relatively constant, and, moreover, there may be no real accumulation of adaptive (phenotypic) change observed for millions of years.

Species, too, are alleged in many cases to remain stable (i.e., component organisms tend not to differ appreciably in phenotypic features through considerable spans of time), particularly in terms of those features unique to the organisms within the species, those in which a species differs from its nearest relative (Eldredge 1971; Eldredge and Gould 1972, 1974; see Eldredge 1985 for a review). Such features may serve economic or reproductive (i.e., Paterson's SMRS) functions, or they may be features with no clearcut or at least determinable significance. It is common to invoke either the homogenizing influence of gene flow, a hypothesis currently much in doubt, or some sort of internal homeostasis (presumably of the developmental pathways of component organisms) as a source of such species stability. No doubt, such processes may be contributing factors. But in most of the cases of prodigious species-level stasis known to me, the species always remain in the same general spectrum of habitat types, forming communities with organisms of the same or closely related species that they always consorted with. Thus, stasis at the species level seems to me largely, if not purely, a function of habitat recognition. After all, abiotic environmental change usually occasions habitat shift or, in the extreme, habitat eradication.[6]

Thus, births, deaths, and persistence (with and without change) of elements (particularly above the organism level) within the genealogical hierarchy—processes intrinsic to that hierarchy—as particular events are nonetheless best seen as reactions to processes and events within the ecological hierarchy (also primarily above the organism level). Below the level of the organism, within the genealogical hierarchy, there may be impetus for change from without (the extrinsic physicochemical inducers of mutation and, of course, natural selection as an effect from a higher level) as well as intrinsic processes. The two process hierarchies of ecology and genealogy are intimately related. I shall now examine some additional aspects of just how these two hierarchies interact.

## Two Basic Models of Hierarchy Interaction

It will be useful at this juncture to describe what I take to be the extremes of possibilities of the manner of interaction among individuals between the genealogical and ecological hierarchies. The two extremes delimit a huge number of possible combinations for how the interactive network may in fact operate. Narrowing of the possibilities to a preferred, smaller set converges on a description of a basic structure of the evolutionary process.

In fig. 7.1, the most complex, outer-limit model is sketched. It shows a hierarchy purist's dream: two parallel hierarchies, each with classes of individuals that are not in correspondence, except, of course, at the organism level, where the two different aspects (economic and reproductive) of organisms are formally allocated to the two different hierarchies. Individuals at adjacent levels within each hierarchy are seen to affect one another in various ways (constraints, boundary conditions, upward and downward causation in general). In addition, there are interactive processes among individuals within levels in the ecological hierarchy (right side of fig. 7.1) and processes of birth and death of individuals at the various levels of the genealogical hierarchy (left side of fig. 7.1). The main feature of fig. 7.1, however, is its illustration of the possibility that each of the sorts of among-hierarchy interaction discussed above takes place directly between individuals of any of the levels of the two hierarchies. Thus, collapse of a regional biotal system—such as the Amazon rain forest—may (potentially) be directly responsible for the extinction of an entire family, say, of freshwater mollusks. The "invention" of a new phylum—say, Arthropoda—may in turn have enormous and direct effects on the composition of the entire world's economic biotic system. A radical mutation might change a key element of the trophic apparatus of an ecosystem, causing its direct reorganization—and so forth.

Plausible as such scenarios seem—and as theoretically conceivable as

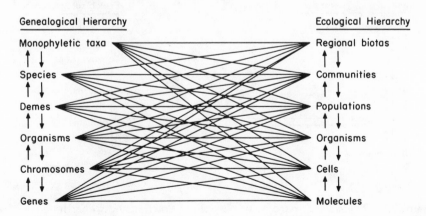

Figure 7.1   One version of the interaction between the ecological and genealogical hierarchies.

they might be given my insistence that entities at all such levels within both hierarchies definitely are to be construed as individuals—nonetheless the shorthand mode of expression in the preceding paragraph is all too obvious. Yes, "invention" of arthropods (and, later, terrestrial insects) has had enormous implications for the very structure and composition of the world's ecosystems. And the extinction of the Phylum Arthropoda, or Class Hexapoda, would indeed have enormous impact on such systems. But just how would these results be effected?

Figure 7.2 illustrates the opposite end of the spectrum of interactive possibilities between the two hierarchies. Organisms are still allocated to both hierarchies to take into account their two very different aspects. The within-hierarchy interactions remain the same, but now the connections between the two hierarchies are seen—as a limit—as purely a function of the identity of organisms as constituent levels in both hierarchies. In other words, under this scheme organisms provide the nexus, the point of entry from one process hierarchy into the other. Natural selection, for example, must work in this fashion. But consider examples of cross-hierarchy higher-level individual interactions—between arthropods, say, and regional biotas. Under the model in fig. 7.2,"invention" or loss of insects is still seen to have enormous effects on regional ecological systems. But here, the way the interaction works is rather more indirect. Hypothetically (but cf. the Alvarez et al. 1980 scenario for Cretaceous extinctions), all local terrestrial ecosystems are adversely affected by some abiotic calamity. Death of organisms is so far-reaching that it occurs within absolutely every ecological system, radically rearranging all such systems. Inasmuch as all such systems are affected, there is also loss of entire species, families, orders—potentially up to and including the entire Class Hexapoda.

The ramifications are enormous. Suppose plants in general were relatively unaffected by the initial catastrophe (actually, in the Cretaceous

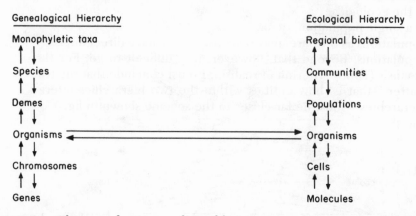

Figure 7.2   The second extreme of possible interactions between the ecological and genealogical hierarchies.

scenario precisely the opposite is alleged to be true). No insects feeding on and fertilizing plants would have enormous ecological implications (and immediate genealogical implications for the plants). No insects— extinction of Class Hexapoda—means that no insects would be available to replenish the ecosystem, to supply additional replacement organisms of the same or similar kind (however vaguely similar they may be—i.e., an insect co-opted from an entirely different family, adaptively convergent on those no longer there). The catastrophe hits all levels of the genealog- ical hierarchy simultaneously, with concordant deaths in all terrestrial ecosystems. And the higher-reaching the effects in the ecological hier- archy, the higher up the genealogical hierarchy the effects are bound to be. The higher those effects in the genealogical hierarchy turn out to reach, the more far-reaching are the ecological consequences when it comes time to rebuild those ecological systems. And all the while, all these events are the simple consequence of the deaths, and births, of organisms. Note that the control of deaths of individuals above the organism level in the genealogical hierarchy by the processes and events at relatively higher levels of the ecological hierarchy carries with it the prediction that the extinctions of most genealogical individuals at any level will occur in phase—at roughly the same time—and will not appear to be scattered randomly through geological time.

This second extreme possibility (fig. 7.2) perhaps runs the risk of revert- ing to what Hull (1980) disparagingly (with justification) called "com- monsense" ontology. The various entities, however, remain as individuals in both hierarchies. Higher-level control of processes remains very much in evidence in this second scheme; it is not as reductive as it may appear at first glance.

Note, too, that these two extremes are purely that—extremes of a spec- trum of possibilities. Direct selection of species, for example, may well be possible should species be shown to possess (1) species-level adaptations (of either basic sort) or (2) some direct, meaningful interactive existence in the ecological realm (beyond the special case of reduction of a species to a single population–deme). Certainly demes, as close in composition to populations as they are, may be construed to have direct interactions with populations. Beyond that, however, it is difficult to go. For the sake of Levins's (1968) criterion of realism, I must conclude that the truth of the matter—that is, how entities within the two hierarchies interact across hierarchies—must lie far closer to the scheme shown in fig. 7.2 than the one shown in fig. 7.1.

## THE FALLOUT: EVENTS AND PATTERNS
## IN EVOLUTIONARY HISTORY

There are a number of rather different definitions of evolution in the lit- erature of evolutionary biology. Some replies to the question "What evolves?" emphasize change in genetic information, others the (genetically based) phenotypic attributes of organisms, whereas others (albeit less commonly) point to the emergence of new species or even to taxa of higher categorical rank. Genealogical evolution in general is simply change in state among $L_{i-1}$ individuals within any $L_i$ individual. All such change is ensconced, ultimately, in the genome, but there are stable levels of packaging of this information, and each level (as Dobzhansky told us) has its different rules of statics and dynamics. And much of the impetus for births and deaths, stasis and change (which amounts to a generalized neutrality versus differential births and deaths via natural selection and concomitant effects) within the genealogical hierarchy stems from the abiotic and integrated abiotic–ecological realm—the ecological hierarchy. One cannot view evolutionary process without a detailed consideration of the economic organization of life.

But what of these entities that evolve, the stable genealogical individuals that variously remain the same or become modified? Harking back once again to Dobzhansky (1937a, 1941), who saw genes and species as two major sources of discreteness in the evolutionary realm, we ask if we can view the entire genealogical hierarchy as a system in which change based on the differences among $L_{i-1}$ individuals through time produces change within an $L_i$ individual through its lifetime (an ontogeny or life history of an $L_i$ individual). Viewed this way, the synthesis is simply a theory that sees (genetic) change among organisms within species, with such change loosely extrapolated to "explain" the entire history of life. What I am suggesting here is that variation, different forms of genetic information, is much more complexly apportioned in the living world: organisms vary within demes, demes (as Wright maintained so long ago) vary within species, and species vary within monophyletic taxa. It is true that all the genetic variation within the Phylum Mollusca reduces to all the genetic information to be found in all its extant classes, which in turn reduces to all molluscan species, thence demes, and finally all molluscan organisms alive at any one moment. But just as Wright (1931) pointed out for demes, how that genetic variance is apportioned in the living world makes a tremendous difference—all the difference in the world—in terms of its fate, its evolution. I am merely suggesting that among-species, within-monophyletic taxa packaging is critical as well, just as critical as demal packaging.

Ernst Mayr's well-known assault on typology affords another way to contrast the synthesis with a hierarchically structured evolutionary theory. At least since *Systematics and the Origin of Species*, Mayr (1942; see also 1982 for a recent statement) has combated the tendency he notes on the part of many systematists to treat species as monolithic units, as if species were not internally variable. As we have seen (chapter 3), Mayr was particularly concerned with contravening this notion because it is this very within-species (and genetically based) variation that affords the raw material on which selection can work to forge and maintain adaptations.

And Mayr was justified in taking this course, if for no other reason than that most species really are polytypic. Moreover, many of the procedures and terms in general use in systematics (including, of course, the various usages of the very word *type*) still come perilously close to a Platonic ideal expressed as if the "essence" (i.e., of the morphological features) of a species are least imperfectly expressed in the single, standard "type" specimen. (That we continue to need reference specimens in systematics is, however, utterly undeniable.) Variation has been treated, to some extent, in the history of systematics more like deviation from the norm than a natural feature of species.

Be that as it may, Mayr's attack on typology has always been curiously shortsighted. One cannot compare any two entities—say, two species of birds—without having some firm idea of the respective properties of those species. It is well and good to claim some appreciation for the range of internal variation of any organismic property under comparison between two species, but in reality Mayr (and every other systematist) distinguishes any two species by their consistent phenotypic differences. Indeed, it was the fact that features variable *within* species seemed to be different from those that varied *among* species that, at least in part, impressed Kleinschmidt and others whom Mayr (1942, p. 114) cites, and which also convinced Goldschmidt (1940) that microevolution and macroevolution must be different processes.

In other words, mental processes alone force one to generalize about the (organismic) properties of demes when comparing demes, or the properties of species when comparing species. After all, it is the differences *among* such entities that interest us in any case, just as it is the differences among organisms that Mayr emphasized as being of material evolutionary significance. But is it all merely a requirement of the comparative process that forces us to make abstract generalizations about the (once again, organismic) attributes of species in order to compare them? Or are there some attributes that are in fact distributed in such a way that they vary among species but not within? Of course there are.

## The Hierarchy of Homology

Darwin (1859) saw that if evolution (his "descent with modification") had in fact occurred there should be a nested pattern of resemblance linking up all members of the biota. Evolutionary novelties—adaptations—would occur only in the original species and, in the same or further modified form, in all of that species' descendants. Put another way, any organismic attribute (i.e., any observable feature of any organism) has a finite distribution in the realm of organisms. As outer limits, a feature may be utterly unique to an organism (as particular fingerprint patterns allegedly are for individual *Homo sapiens*) or may be found in absolutely all organisms that have ever lived (as RNA appears to be). Most organismic attributes are distributed in the vast range between these two extremes. This observation, conforming so completely to theoretical expectation, is the strongest corroboration of the general notion of evolution. It is, as well, the cornerstone of cladistics, as the notion that all organismic attributes have a finite distribution dictates the task of systematics to be the analysis of those very distributions.

A peculiar sort of hierarchy instantly emerges from these overlapping distributions of evolutionary novelties, the vast majority of which were probably adaptations in the purest sense (Gould and Vrba 1982) at the point of their introduction ("invention") in evolutionary history. Some features are more generally distributed than others; hair is more widely distributed than carnassial teeth in mammals, for example.

But there is another more easily perceived aspect of hierarchy involved with organismic attributes. Homology itself is a hierarchical concept; the labyrinthine conceptual history of homology aside, homologous characters are simply the various known conditions that a phenotypic feature has assumed in the course of evolutionary history. Thus, the hair of mammals and the feathers of birds are said to be homologous with the flat ectodermal features ("scales") of early amniote tetrapods. Tradition has it that scales are one state, hair another, and feathers yet another state. But it is probably more accurate to see scales as a more generalized (undifferentiated, primitive; the terms are by no means synonymous but all come to mind) condition, while both hair and feathers are merely derived subsets of the more general condition "scales." Viewed this way, homology is clearly a hierarchical concept. And it is technically true, though epistemologically probably useless, to acknowledge that the general condition of the soma is that of a single cell (with an undifferentiated nucleus at that), so that in a fundamental sense (although it is almost impossible to visualize) all evolutionary novelties are homologous in a single, yet complex, hierarchical fashion. It is more useful, I think, to view homologies as hierarchical systems of organismic variation—the products of transformations of adaptations. And adaptations, distributed in varying degrees throughout the living world, themselves seem ranked in terms of their breadth of distribution.

### The Linnaean Hierarchy

The taxa of the Linnaean hierarchy are classically nested inclusive series of individuals ranked according to a linear list of categories (classes). All the taxa in the real world—that is, not necessarily those defined and recognized by systematists—are "real" individuals; they are actual skeins of species united by ancestry and descent, including an ancestral species and all its descendants.

And yet the particular monophyletic taxa so named are clearly just the branches of the genealogical "tree of life," and not each and every bifurcation need be, or ever is, named. All taxa that are defined and recognized are so designated on the basis of sound genealogical principles, recognition of the joint possession of some unique feature that stamps all members as descendants of a single common ancestor. In deciphering the history of manuscripts, it is the commission and transmission of copying errors that supplies the pattern of novelty (diversity) and similarity (Platnick and Cameron 1977). In languages, it is the invention of new words, new grammatical structures, and ultimately still more general features of languages. In phylogeny, it is the forging of new adaptations, their transmission, and their subsequent modification.

It is a fact that, whereas it is difficult to conceptualize the distribution of organismic attributes as forming a single coherent hierarchy, it is those attributes that define monophyletic taxa, and those taxa are nested into one grand hierarchical array (which, as Darwin saw, corroborated the very notion of evolution). We define and recognize such taxa on the basis of these synapomorphies, yet the taxa seem more coherent, more tangibly real than these complexly distributed homologies that can be so difficult to analyze. Linnaeus said that "the taxon gives the characters," though methodologically we proceed the other way around. Enter, once again, ontology: monophyletic taxa are individuals, more than just metaphorical branches on the phylogenetic tree. Organismic attributes are no less real as features on a particular organism: they become classes or, at the very least, classlike when we speak of more than one organism. We are tracing the history of the information of the genome, in some cases as it variably changed or remained the same through as many as three and a half billion years. It's a small wonder that taxa seem more concrete. Again we depart somewhat from the point of view typified by Dawkins (1982), whose perspective on this extended version of his Necker cube tends to see concreteness and permanence the other way around, ensconced in the genome.

## Phenomena of Evolutionary History

At any point in the history of life, one can take a freeze-frame photograph and observe a single hierarchical array of organisms, species, and monophyletic taxa, linked together by a complex distribution of organismic features. These twin, concordant hierarchical schemes are usefully regarded as historical patterns. They are the results of past interactions between the ecological and genealogical hierarchies. In addition, the genealogical array present at any moment continues to supply the organisms (replete with the requisite economic and reproductive adaptations) to keep the current ecological arena functioning, which in turn changes the complexion of the genealogical array by the time we take the next snapshot. A pragmatic consequence, of course, is that we can study the biota extant at any one point in time—the Present springs to mind—and usefully claim that we are in fact studying the history of life. From this perspective, inclusion of the fossil record merely adds to our appreciation of the total diversity of life and to our analysis of the historical distribution of adaptations. That the entire history of life, from a genealogical, "informational" point of view, is hierarchically arranged is a simple consequence of the functional arrangements of the living genealogical (process) hierarchy itself and comes as no real surprise. We still need to know why we have these hierarchically arrayed individuals in the first place—a point to which I return below.

For much the same reasons, it is no surprise that there is an ecological history of life. The various extant individuals of the ecological hierarchy, as labile (or at least indefinitely bounded) as they seem to some ecologists, all have histories. The larger units seem to change less rapidly than the smaller ones. The native Australian biotal system is quite old, even if the particular species of eucalypts and marsupials, to single out but two of the endemics, are not the same as those that founded the system. There is hierarchy in the ecological history of life just as there is in the everyday economic integration of the biosphere.

Yet there is a linear component to the genealogical and ecological histories of life—the familiar linearity common to most historical analyses, regardless of the nature of the system. We start at the beginning and work up. Whether it is the history of a nation, the cultural history of *Homo sapiens*, the history of gene frequencies within an experimental population, or the history of the Primates, the goal (as Teggart 1925 put it) was to understand "how things have come to be as they are today." The usual procedure is indeed to start with a description of a system at some chosen "beginning" and to describe the changes through time—as seen in an embryo at various stages, as sampled in a laboratory culture of *Drosophila melanogaster*, or as described by sequential sampling of fossils up the geological column. The fossil record is a particularly effective source of information on the linear ecological history of life—the sequences of associations among organisms. Indeed, patterns of cross-genealogical extinction—and origination—are one of the two main empirical generalizations of the fossil record, generalizations of deep evolutionary interest. (The other is patterns of stasis and change within individuals of the genealogical hierarchy above the organism level. Neither pattern is adequately considered by the synthesis.)

Yet linear description of history is fraught with potential error. It is, in fact, practically an impossible task. All such systems require a formalization of procedural rules—a detailed epistemology—for their analysis. For the genealogical history of life, such rules appear to be available already (Hennig 1966; Eldredge and Cracraft 1980; Wiley 1981; see Nelson and Platnick 1981 for a still more general approach). I imagine that a similar analysis of inclusive hierarchical groups would be required for a careful reconstruction of the ecological history of life—a method that so far seems uninvented, though perhaps partly approached by "vicariance biogeography" (Nelson and Platnick 1981). Historical analysis of stasis and change within lower levels of both the genealogical and ecological hierarchies are perhaps less problematic, dependent as they are on sampling of the systems at various stages of development; events in such systems (e.g., developing organisms, selection experiments, the development of a sere after environmental downgrading) take place within the professional lifetimes of individual biologists, greatly facilitating at least the completeness of description of the statics and dynamics of the system. But even these systems are not without their problems of observation and analysis.

The "ideographic" description of the history of genealogical and ecological systems, interesting as it may be in its own right, reveals information on the nature of the evolutionary process itself only when we compare theoretical notions of process with that history. And here, it seems to me, individual events ("invention" of the echinoderms, the terminal Cretaceous extinction, the changes in allelic frequencies in a laboratory population of house mice) are useful only to the extent that we can generalize on them, which means to the extent that such events fall into repetitive groups, that is, classes or familiar patterns. It is the comparison of these patterns (classes of evolutionary events within the various levels of both hierarchies) that renders any aspect of a process theory of evolution testable.

The classes of events considered by the synthesis in detail are relatively few in number (see Chapters 2 and 3 and a summary statement in Chapter 4, p. 94). A more formal consideration of a rather more broad listing of patterns of evolutionary relevance cannot help but increase the scope of a theory of the evolutionary process. In general, processes within both hierarchies that center around organisms are those that have been most intensively studied and are, consequently, probably the best understood. Patterns involving the higher-level individuals within both hierarchies— as well as those at lower (molecular) levels—have received less attention and are, at the very least, far less completely understood. The ontology of these entities either has not existed (molecules) or has been confused (higher-level individuals). The synthesis dealt with these entities as a black box (justifiably) in the case of the lower-level entities; the upper-level entities were treated as a cumulative outgrowth of the better understood processes at the among-organism, within-deme (or simply within-species) level.

One response to such treatment of higher-level phenomena in recent years has been to assert that there are processes intrinsic to the higher levels that themselves account for patterns of stasis and change within those upper levels. And, indeed, there are; processes of birth, at least, are seen to be different for such individuals at higher levels (demes, species, monophyletic taxa) from what they are for organisms. But other processes, such as group-level (i.e., $L_i$ individuals) selection have been invoked as well—and largely in vain, as I have earlier argued.

A revamped ontology of higher-level individuals within both the ecological and genealogical hierarchies at once reveals the narrowness of the scope of the synthesis and outlines the structure of a process theory of evolution. What emerges is not a parallel system of stasis and change with reference to "new" processes unique to each level, as has been stated or claimed in some of the earlier enthusiastic expositions of hierarchy (including some of my own; cf. Eldredge 1982b, in part). What really seems to be going on in the upper-level individuals of both hierarchies is consequent to no small degree upon what we think we know routinely transpires among organisms within both populations and demes. But the significance of those lower-level processes—the way they work in these communities, regional biotas, demes, species, and monophyletic taxa—is quite different from the conception of the synthesis.

A major paradox confronting contemporary evolutionary biology is that the data of systematics and paleontology strongly suggest that species are being sorted, just as Wright has suggested that demes undergo a sort of differential birth and death. The patterns of histories of species, when simply described, show precisely this. There is definitely a differential production, and perhaps especially survival of hominid species during the last four million years—a pattern not well described by an image of a shifting adaptive landscape, with hominids gradually increasing in brain size as an adaptive response to the selective need for greater intelligence or progressively more complex social organization. Such patterns, so very plain to paleontologists, seemed naturally to require a mechanism—hence the plethora of species selection arguments since 1972, when Gould and I suggested the necessity of such a process to explain such patterns as directional evolutionary change that appeared to proceed by interspecific patterns of origin and extinction, always granting that the change at speciation and the origin and modification of the adaptations were under the control of "conventional" natural selection.

Now, if the notions of species selection are falsified by the observations that (1) economic adaptations, at least, solely involve organismic attributes, not species-level properties, rendering species selection meaningless (my interpretation of Vrba's point—Vrba 1983; Vrba and Eldredge 1984), and (2) in any case, species do not seem to be active interactors in the ecological arena, then *so too is the entire adaptive landscape metaphor of adaptive macroevolution simultaneously falsified*—provided, of course, that species are construed as individuals and the literal truth of the two observations above is agreed upon. The extrapolationist, macroevolutionary portion of the synthesis is invalid because such entities have been seen as classlike collectives of similarly adapted organisms, sharing a common adaptive history and destiny. That is, to the extent that they are seen as entities, they are construed primarily as *economic* entities. That leaves the synthesis with nothing really more than the core neo-Darwinian paradigm of natural selection and drift moderating stasis and change in allelic frequencies within populations–demes on a generation-by-generation basis—a paradigm that few have trouble seeing as a dynamic of nature, but a paradigm that itself is hardly a "synthesis" except in the narrow sense of a merger (achieved by Fisher, Haldane, and Wright) of what was then known of genetics with the vision we have all inherited from Charles

Darwin. Not for nothing did Gould (1980a) speak of the demise of the synthesis.

Having recapitulated in the last several paragraphs some of the main arguments of this book, it remains for me to characterize the general nature of what I take to be inadequately addressed evolutionary patterns—classes of evolutionary events—and to show how a hierarchically structured theory can handle them. I shall borrow yet another page from Dobzhansky (1937a) and sketch a general outline of the entities, patterns, and processes that make up this thing we call evolution.

## TOWARD A GENERAL EVOLUTIONARY THEORY

In chapter 4 I presented a comparison of Dobzhansky's (1937a, 1941), Mayr's (1942), and Simpson's (1944) early theories, and in so doing sketched an outline of the early synthesis in terms of its ontology and its views on evolutionary patterns and processes. I shall now attempt the equivalent treatment for my (admittedly idiosyncratic) version of a hierarchical evolutionary theory. I have presented detailed arguments and explorations on ontological matters, plus notions of process within and among levels and between hierarchies. I have had less to say up to this point on specific evolutionary patterns, which require particular concatenations of process for their full understanding (as Simpson pointed out in 1944). Nonetheless, in this summary section I will refer to specific patterns and the general nature of their production by evolutionary processes a bit more fully.

Why do all these various sorts of individuals of the ecological and genealogical hierarchies exist? They are stable entities. I will start only with some form of ur-organism, which minimally possesses a strand of DNA and some inclusive packaging that protects as well as procures enough energy from the external milieu to at least grant the possibility of replication.[7] Thus, we start with a simple soma and genome, which is to say an organism with a genome used both for the production of economically important products and for reproduction. Such an organism, many times less complex than even a bacterium, nonetheless has all the basic requisites of any organism in the two hierarchies. It was only later (much later, really, nearly three billion years after the "invention" of life) that the somatic and germ lines really did become separate, with the invention of multicellularity. There is little physical separation, even if the two functions are distinguishable, into separate structures for matter–energy transfer and reproduction in prokaryotes or in unicellular organisms in general. Dawkins's acknowledged perplexity over why there should be organisms at all is itself mystifying; something has to go to work *and* keep house if those precious genes are to be able to do their job.

Elucidating the "purpose" of things—in this case, of these various sorts of biological entities—is a tricky business. Ghiselin (1974b) is especially eloquent on the modern version of teleology in evolutionary biology, which forms a kind of "hidden agenda." When we ask "Why do we have organisms? Why do we have sex?" what we mean is usually "Who or what benefits from this arrangement?" At the risk of sounding ultra-materialist, most of my answers to such questions take this form: "Given the prior condition A, B is a simple and necessary consequence." For example, I have just implied that organisms really exist because genes require protective and energy-procuring packages. And organisms reproduce. So haven't I agreed completely with Dawkins after all that an organism is the genome's way of making another genome? Well, yes, but it is equally true (and merely part of Dawkins's Necker cube) that a genome is an organism's indispensable tool for making another organism. There is no chicken-and-egg dilemma here, for both, in rudimentary form, were there together from the Beginning as a matter of necessity.

Skipping demes, only for the moment (I agree with Dobzhansky and virtually everyone else that demes are difficult both to characterize and to "experience"), we come to Dobzhansky's next major discrete entity: species. Why are there species? There is no doubt that species are stable entities and thus prime candidates for a general level, a slot in the hierarchy. Recall Dobzhansky's imaginative response to the problem: species are reproductively isolated units on an adaptive peak. It is their reproductive isolation that keeps them on that peak; otherwise, inharmonious gene combinations would form through hybridization, and there would be no way for groups of organisms to occupy adaptive peaks, let alone to focus on perfecting adaptations, to adapt to the changing positions of peaks, or to scale new peaks. Dobzhansky saw the existence of reproductively isolated species as adaptive and strove mightily to formulate a theory of the attainment of reproductive isolation through natural selection. In fairness, it must be stated that the synthesis never fully adopted this particular set of Dobzhansky's views.

Enter the work of H. E. H. Paterson, whose notion of specific mate recognition systems (SMRS) puts the problem the other way around. Specific mate recognition systems are what I have otherwise been calling reproductive adaptations. Species are isolated from one another as a simple consequence of the continued necessity for sexual organisms to recognize one another. As they continue to do so, the adaptive complex that allows reproduction to occur is disrupted occasionally by the accidental establishment of geographic isolation—an ecological phenomenon. Apparently, the broader the SMRS, the harder it is to disrupt. Thus, speciation in sexually reproducing organisms (a redundancy—only sexual organisms form species) is *not* a process of adaptation. It is, as Paterson and Vrba have stressed, the disruption of reproductive adaptation, a side effect of isolation and the continued need for organisms to continue to mate when in isolation. Minimally, only reproductive adaptations need be involved in speciation; alternatively, reproductive isolation may arise as a side effect of divergent economic adaptations in isolated populations. The formation of new species is an accident and not, as Paterson points out, a direct product of adaptation, of selection for reproductive isolation. Speciation is the simple consequence of sexual organisms living in a heterogeneous, shifting environment.

Which leaves the enormous side issue: Why is there sex? Again it is useful to remember Dobzhansky, who saw the tradeoff between the virtues and pitfalls of a species concentrating its (organismic, economic) adaptations too narrowly, too close to the adaptive peak. Selection would seem logically to favor such a narrowing, but only at the sacrifice of the plasticity, the flexibility that retention of a goodly amount of genetic variation affords against the (inevitable) change in position of the adaptive peak. Sex, to Dobzhansky, ensured at least some measure of variability. At least since Muller (1932), geneticists have tended to see sex as affording the opportunity for rapid evolution. Later theorists, Williams foremost, scoffed at such notions because selection can have no eyes for the future; there can be no saving for a rainy day, no selection "for the good of the species."

It turns out that some species are generalists while some are specialists in terms of their economic adaptations, and these "niche strategies" are, of course, based upon the anatomical, behavioral, and physiological properties of individual organisms. The explanation for why some organisms, hence species, are overall generalists (eurytopic) while others are specialists (stenotopic) has to do with different economic "perceptions" of the environment (and not the "future"—Williams is right). Why the organisms of some species remain eurytopic while those of other clades remain stenotopic remains a difficult problem, though not (thankfully) the central issue here (see Eldredge 1979, Eldredge and Cracraft 1980, and Vrba 1980 for further discussion).

Ghiselin (1974b), Williams (1975), and Maynard Smith (1978) have all written books in the last decade that explicitly address, among other topics, the basic problem of why we have sex—and why sexual reproduction predominates. Stanley (1975b) has addressed the problem from a viewpoint similar to the one I take here, as have Fowler and MacMahon (1982). Stanley's review (1979, chapter 8) is particularly useful from a hierarchical standpoint. And Hull (1980) lists the prevalence of sex as one of the great outstanding conundra of contemporary evolutionary biology.

Williams (1975) and Maynard Smith (1978) both assume that sex must be *for* something. Williams strives to develop models that demonstrate the selective value of sex for the participating organisms (despite the fifty percent "cost" of meiosis). But in his final chapter he paints a picture rather close to the views I adopt here. According to Williams (1975), the main effect of sex in "biotic" evolution is *not* its retention of flexibility to provide the wherewithal for rapid evolution. It is instead the greater resistance to extinction that less narrowly focused genetic systems afford that simply adds up to the observed predominance.

Stanley (1979, chapter 8) takes a similar approach. But is this argument sufficient to explain the predominance of sex? Maynard Smith (1978), while outwardly more receptive than Williams to such a group selection argument, is by no means as optimistic on such long-range possibilities that sex may offer.

And indeed, given that sex is the one physiological function not required for the survival of an organism, there is a bit of a paradox here. Why should an organism even bother reproducing? What is it that leads to that urge to mate—is it Dawkins's selfish genes? Why does Williams say that the goal of the fox is to leave more genes to the next generation, while we contemplate a fox dining on (perhaps even enjoying) a quail? What can an organism or a gene care—even metaphorically—about being represented in the next generation? It has to be a matter of complete (functional, biological) indifference to an organism, human consciousness completely aside. Reproduction—of any sort, asexual as well as sexual— is achieved at a cost.

So why do we have reproduction? There can only be one answer: only those entities that engaged in reproduction are still around. Life would have become extinct long ago were there no reproduction. That is the "trivial" truth. Protoplasm is mortal; were it not, I believe, we would not have a hierarchy (two, actually) of biotic elements in the real world. Or at least it would be a vastly reduced hierarchy. Life got going minimally three and a half billion years ago. The only imaginable form of life that could have survived that long—given the way of all flesh—is that which makes more of itself. Now *that* is a gross form of selection.

Given the prevalence of reproduction, why the prevalence of sex? I am not competent even to speculate why sex was "invented." It seems not so much a matter of advantage as of cellular invasion, otherwise so common in the early history of life if Margulis's (e.g., 1974) generally compelling notions of symbiosis are to be heeded. Loosely speaking, we may point to "heterotic" effects (in the early stages of life, for protection against damage or loss of genetic information, as Bernstein et al., in press, argue). The question that I feel is virtually automatically answered with reference to hierarchical systems is why sex is the *prevalent* form of reproduction.

The answer, again, seems obvious: sexual organisms form stable groups—species—that commonly last for millions of years. That's right; the argument is superficially totally circular. We have species because we have sexually reproducing organisms. Species are a simple and necessary consequence of sexual reproduction. But species are stable entities. They tend to persist. They send colonies into different habitats, typically retaining membership in more than one community, more than one ecosystem. Species are powerful, conservative entities. They are difficult, in general, to get rid of—as witness the multimillions of years' duration that marine invertebrate species recognizably display. Difficult to dislodge, they easily outmatch clones and parthenogenetic lineages. They do, as Dobzhansky said, retain a genetic flexibility (through recombination as well as polytypic adaptive variation), a genetic flexibility that is very potent, especially when recruitment to new (or downgraded) habitats is in question, rather than the "iffy" scenario—"What if the environment should change?"—Dobzhansky had in mind. The flexibility is entirely a consequence of sexual reproduction. The point is, once it is there it has enormous implications. The systems works, it is self-perpetuating (as long as there continues to be an influx of solar energy and as long as no cross-genealogical extinction event is one hundred percent effective), and sexually based species simply overwhelm asexual organisms in numbers. As a side effect of sexuality, sexual species are better able to resist extinction than asexual lines. Indeed, strict asexuality (as opposed to some form of alteration of generations) is truly rare and, apparently, always represents a secondary loss of sexuality. In such cases we need an ad hoc selection argument, and such would be, in principle, straightforward, as there is that "inexplicable" cost attached to sexual reproduction.

This line of argument on the very existence of species differs, of course, rather markedly from Dobzhansky's explanations of the discontinuities between species: discrete species keep organisms perched squarely atop adaptive peaks. Reproductive isolation, in his view, prevents the formation of unharmonious gene combinations. Species (reproductively isolated groups) are a means to focus economic, organismic adaptations. Bernstein et al. (in press) have recently argued that the mere exigencies of sexual reproduction produce "distinct and relatively homogeneous groupings of individuals"—that is, species. Bernstein et al. stress the unity of species (a theme variable within Mayr's writings over the years), whereas it is their internally variable nature (with respect to the economic adaptations of component organisms distributed geographically) that seems implicated with species stability (as its effect is to reduce the probability of extinction). In Dobzhansky's and Mayr's early writings, as we have seen, the importance of within-species variation in economically significant aspects of organismal phenotypes was the promise the variation held for future adaptive modifications of those features.

So much for why we have species. The consequences for having species (beyond the concomitant prevalent retention of sexual reproduction) are something else again. Species are the means by which sexually reproducing organisms make more organisms, but the ongoing production of new organisms is also a species' (or a deme's, of course) way of keeping going—Dawkins's Necker cube up one step.[8]

Retention of sexuality in the face of the alleged fifty percent "cost" of meiosis may truly be a paradox from a strict selectionist standpoint and if we begin with the premise that every phenomenon we observe is "for" some purpose or another. The scheme I am sketching here sees the components of the genealogical hierarchy in a purely mechanistic sense. Nothing exists for a purpose. Organisms procure energy no more exclusively for reproduction than for just staying alive. Sex has automatic consequences; it is no more for the benefit of the clade (as Stanley 1979, p. 215, argues) than it is for the benefit of the organism. The paradox of sex might

not be totally dispelled, but the answer appears not to reside in the traditional sort of optimality or tradeoff arguments so common in the selectionist literature.

I have emphasized in chapters 5 and 6 the nature of species as conservative entities. It would be as well to reiterate at this juncture that the
properties I am discussing are those directly palpable phenotypic features
of individual organisms: their size, color, conformation, anatomies, behaviors, physiologies. I am only referring to such phenotypic properties that
appeared under the aegis of natural selection within demes. I agree with
Dobzhansky (1937a, 1941) and Mayr (1942) and all later careful authors
that not all phenotypic organismic attributes are, or were originally, adaptive—yet surely most are, or were, in some sense, even if they have since
been co-opted for still other functions (the "exaptations" of Gould and
Vrba 1982). And I repeat what really so far remains the one grand generalization of evolutionary biology, one we owe to Darwin (1859): adaptations are forged, maintained, and modified by the differential perpetuation of heritable variation. This pattern–process, this bias in what is
perpetuated and what is not, is the generation-by-generation relative success in the reproductive activities of organisms within demes. The process
is alleged to be creative, but, as Mayr has said, selection is a "two-step
process," and the actual creativity may as justifiably be seen to reside in
the spontaneous generation of a change in the genetic information and its
presentation in the phenotype. As Wright (1945) remarked, there is nothing more or less creative at the second step—the "keep what works, get
rid of what doesn't" step—than selection at any other level. So much is
well known but always bears repeating, as Mayr says.

But this culling, this selection of *organismic* traits, is not confined to
generation-by-generation selection within demes. This is where there has
been a tremendous amount of confusion. Organismic phenotypic traits—
adaptations—are treated at the organismic level; they are seen to persist
at the deme (population) level in conventional theory. But they also exist
at the species level, and species are strung together into monophyletic
skeins—skeins we recognize through the analysis of the very distributions
of such phenotypic traits. The packaging of *organismic* adaptations does
not stop at the population level (including demal, depending upon what
sort of adaptations are involved). Species and monophyletic taxa are also
such packages, and the Necker cube holds once again: a monophyletic
taxon is but a species' way of making another species (literally true), but
the ongoing production of new species is how the stable entities we call
monophyletic taxa stay alive.

Thus, there is differential species survival, as well as differential species
births. To the extent that such is deterministically caused, it is based not
on any imagined properties of species per se—so-called emergent properties of species. It is based on the *species-wide distribution of organismic
attributes*. Hence, in the purest sense, it cannot be species selection. Thus
argued, traditionalists (e.g., Williams 1966) tend to brush the phenome-

non aside as trivial. But the comings and goings of species, the redistribution of what organismic attributes are actually present at any one moment, and the consequent configuration of ecological units of all magnitudes form nothing less than the dominant theme of the history of life. *Patterns of species origination, existence, and extinction and the corresponding patterns of stasis and change of phenotypic features remain wholly unaddressed by the synthesis.*

Thus, to the extent that it is deterministic—that is, based upon the workability of a particular set of phenotypic features—there is a gross form of selection at the deme, species, and even monophyletic taxon levels. Call it sorting, not selection; selection connotes properties intrinsic to that level (Vrba 1984b; Vrba and Eldredge 1984). In ontological terms, note that species sorting, working as it does on organismic attributes, focuses on classlike attributes of species. The trick is to remember that species are also individuals because of the reproductive processes of sexual organisms. Fowler and MacMahon (1982), ecologists who have seen the importance of the process, refer to it as "selective extinction." How does it work?

The regulation of both stasis and change in the general hierarchy model comes from the ecological hierarchy, from the dynamics of integration of organisms with other organisms and with the abiotic realm. Here what works and what doesn't are determined (except for purely reproductive phenotypic features). Recall, too, that it is the second of the two extreme possibilities of interhierarchy interaction, where organisms (almost exclusively) form the nexus, the point of entry, between the two hierarchical systems. Thus, retention and elimination of genetic information as expressed in organismic phenotypes within individuals at higher levels of the genealogical hierarchy are, of necessity, phenomena of lower-level causation. Species live or die as their organisms live and reproduce, or fail to do so. Species, as units, are not direct ecological interactors. Here we contact Vrba's (1980, 1983, 1984b) valuable contribution toward understanding how the evolutionary process really does work: her "effect hypothesis."

The existence of demes, species, and monophyletic taxa as stable entities flows as a simple consequence—an effect—of the basic organization of living systems. In much the same way, we can understand differential births and persistences of species as statistically deterministic outcomes of the phenotypic properties of their component organisms. As an example, consider eurytopy (Vrba 1980, 1984a); it has often been invoked to understand relative ability to evade extinction, and indeed there is some evidence that species in generally eurytopic lineages (i.e., lineages composed of species, in turn composed of eurytopic organisms) are the main contributors to the ranks of "living fossils" (see the recent compilation in Eldredge and Stanley 1984). But consider, as Vrba does as an example, the differential probabilities of speciation: one common pattern in nature is that clades of eurytopes speciate at far lower rates than those composed of relatively more stenotopic lineages (see the papers by Novacek, Vrba, and others in Eldredge and Stanley 1984 for well-documented examples). Vrba postulates that organisms that are eurytopic with respect to their economic adaptations tend to be eurytopic as well in their reproductive adaptations (SMRS). There is some evidence for this, for example, in the horseshoe crab *Limulus polyphemus* (Cohen and Brockmann 1983; Vrba 1984a). And broadly based, flexible reproductive adaptations are simply more difficult to disrupt—a simple hypothesis that explains differential rates of speciation based purely on the phenotypic properties of organisms. The effect hypothesis, however, is a far more general statement; actually, the way the ecological hierarchy affects both stasis and change in genetic information at the deme, species, and monophyletic taxon levels of individuals within the genealogical hierarchy comes through Vrba's effect hypothesis.

But is there no control from above? Of course there is. We are accustomed to paying at least lip service to constraints in evolution: new adaptations must be based on the phenotypic properties—and their genetic basis—that already exist, plus whatever "new" genetic variation may become available. Usually seen as lower-level constraint arising from the genes, such a severe restriction on the "realization of the possible" is perhaps best seen as constraint from above. What is possible depends upon what has already evolved, and particularly on what has survived to any given moment. Enter monophyletic taxa as the storehouses of that information. Suppose, for example, that atmospheric oxygen tension rises to the point where it is feasible for animal life to exploit the subaerial terrestrial environment. Looking around for the variation that might serve as a preliminary working version of the requisite anatomy, physiology, and behavior, one can easily imagine that the problem is not so much one of picking out suitable variants within a deme but rather *among* species, and even more realistically among certain lineages, such as Chelicerata, which early on seem to have had some experience in the subaerial littoral. Only a handful of the scores of marine phyla ever made it to shore as permanent residents. But to forge such adaptations from preexisting variation, all we need is natural selection—among organisms within demes.

And the higher levels have a still deeper impact than mere definition of the limits of the possible. The helpless feeling so common in human life, where the course of human events (political, usually) is literally far beyond anyone's personal control, is nicely mirrored in nature. Events in the economic arena, vicissitudes most easily seen in the abiotic realm but subtly there as well in purely biotic constructions (communities), have a way of sweeping much before them. Regional biotal crashes have cascading effects; the more far-reaching the ecological effect, the more disruptive and far-reaching the effect up the genealogical hierarchy. Organisms are simply lost in the shuffle. Sudden impact of gigantic meteors, with its sudden deaths for organisms of many economic systems and genealogical affinities, is not a good example, for the simple deaths of organisms have their ramifications up both hierarchies instantly and simultaneously. But consider the loss of a key trophic element of an ecosystem. Should that system decline and crash as a result, we have a clear example of a high-level control *in the ecological hierarchy* radically reshaping the earth's inhabitants, thus radically affecting the possible—what will happen in future evolution.

## APPLICATIONS

I said at the outset (chapter 1) that the value of an expanded, hierarchical picture of biotic nature, and specifically the worth of a hierarchically based evolutionary theory over the restricted vision of the synthesis, would be most obvious if old, persistent problems yielded to insights from the newer system and (particularly) if the range of problems relevant to evolutionary theory was to be expanded from a hierarchical perspective.

As for old problems, the perspective does indeed change with the advent of hierarchical levels, with its concomitant recognition that information and matter–energy conservation and transfer are equally important at every level of biological organization. Hull's list of three outstanding problems (the prevalence of sex; the ubiquity of within-species, inter-organismic variation; and multilevel selection) yields as follows: sexual reproduction can, under certain circumstances, be seen to benefit organisms (via heterotic effects and a number of additional models promulgated nearly every month in the technical literature). The adaptive advantages of sexual reproduction at its inception are possibly also attributable to het-erosis or as a way of "dealing with damage or loss of genetic information" (cf. Bernstein et al., in press). Yet a more generally promising approach eschews purposeful explanation and looks at the consequence of sex once it has been "invented": sexual reproduction results in communities of reproductively interactive organisms, entities that remain stable per se but are susceptible to fragmentation. Asking why asexuality is not more prevalent (because asexual organisms can focus economic adaptations more precisely, or because they leave a more exact representation of their genes) utterly ignores the facts that (1) many allelic forms really seem selectively neutral, and (2) in any case, narrow niche breadth (construed in terms of the physiology, behavior, and morphology or component organisms), typically commingled in the synthesis with genetic homozy-gosity, is not, on the empirical face of it, necessarily to be construed as a desideratum in nature. Eurytopes may speciate less frequently but typi-cally (as species) last longer and ordinarily produce as much biomass as their entire sister clade of stenotopes (see Greenacre and Vrba 1984 for relevant statistics pertaining to antelopes). In any case, the relationship between genetic heterozygosity and range of phenotypic (in the broadest sense) properties is terribly vague. At base, variation is maintained at higher levels because sex prevails, as long as the expression of that varia-tion (1) is neutral or (2) contributes to eurytopy.

So why does sex prevail? Beyond heterosis, it must be because the existence of sex implies the existence of species as a consequence. And sexual reproduction implies recombination, implying variation. And species are insidious. With rare exceptions, a species is integrated into far more than a single community in nature. Empirically (as judged, ironically enough, in terms primarily of the economic adaptations of their component organisms), species truly are stable entities. Sex prevails at least in large measure because it creates stable, extinction-resistant entities in nature.

Regarding selection at higher levels, I have argued that, given Hull's (1980) specification of interactors and replicators both being necessary for any form of selection to occur, selection simply cannot occur at any level higher than organisms in biotic nature. Specifically, I have argued that species are reproductive entities, concerned strictly with the conservation and transmission of information (even though that information concerns economics far more than it does reproductive matters). Species, in short, are not meaningfully to be seen as economic units in nature. This, coupled with the observation that no one, at least as yet, has proposed plausible *species-level* economic adaptations, makes species replicators (as Alexander and Borgia, 1978, say, they do make more of themselves) but not interactors. From such a perspective, neither demes, species, nor monophyletic taxa can be selected. (Demes do converge on identity with populations, however, and as a limit are candidates for true selection.) Nor can higher-level ecological units be selected; they are interactors but not replicators. Yet the interaction between these higher-level units within each hierarchy ends up in a pattern of sorting of higher-level genealogical individuals, a pattern that mimics selection yet recognizes that the economic and (genetically based) informational twin functions of life are segregated into two parallel hierarchies above the level of organisms.

So much for old problems. The power of hierarchy is to expand the range of insight into old problems and to recognize as well that the neo-Darwinian paradigm is a potent source of explanation for many of life's evolutionary conundra—as far as these have been admitted into the arena of explanation. There remain some additional topics, such as the main events in the history of life. Beyond genetic hand-waving, there simply is no way of grappling with *the actual events in the history of life* (cf. Teggart 1925) unless we look at the events from the sort of hierarchical perspective that I advocate here. The following is a case in point.

Trends are a classic example, because here, if anywhere, the synthesis simply says that what goes on in a *Drosophila* population cage suffices in an extrapolated mode to explain major patterns in the history of life. Yet nowhere is there a clearer need for a hierarchical perspective—one that admits that there are all manner of biological units, economic as well as genealogic—than in the struggle to explain *real* evolutionary trends.

K. S. W. Campbell (1967) has described such a trend in phacopid trilobites from the Lower Silurian up through the Middle Devonian. The time span is some sixty million years. The biogeographic arena is essentially worldwide. However, there was considerable continental movement over that temporal interval; ecologically, the organisms are shallow-water epicontinental marine, and not all continental areas were continuously inundated during the interval.

Given these caveats, the record of the subfamily Phacopinae during this time span is remarkably strong. The subfamily is unquestionably monophyletic.[9] Campbell (1967) describes what can only be construed as a rather modest series of anatomical changes during the course of the entire known existence of this subfamily. Of the changes Campbell documents in several anatomical systems, I shall focus on but one: the progressive modification of the nature of the surface "ornament" (prosopon) characteristic especially of the cephalon (head) but also of the thorax and pygidium (tail). This and Campbell's (1967) other trends, incidentally, are easily verified with reference to original specimens and to the primary literature; in my experience with these trilobites, I find Campbell's basic characterization of the sequence of phenotypic change quite accurate.

In brief, the predominant diffuse and finely granular ornament of the exoskeleton changes in a linear, progressive manner from the Lower Silurian species of the primitive genus *Acernaspis* gradationally through granules aggregated into small tubercles and covered with a groundmass of granules in *Ananaspis* (Middle and Upper Silurian), to granules covering somewhat larger tubercles in the Lower Devonian *Paciphacops*, to the unornamented, large tubercles of the Middle Devonian *Viaphacops* and *Phacops*. I reproduce here Campbell's (1977) interpretation of the phylogeny of these (and related) genera as fig. 7.3. According to his diagram (and the synthesis), there is no problem in interpreting this trend (and other correlated anatomical trends) within this lineage of Phacopinae as linear, gradual, and presumably adaptive (though the functional significance of the structures and their modifications remains unclear). The reader should bear in mind that this is no radical transformation of anatomical configuration; the example here is indeed modest, hardly the stuff of macroevolution. Yet the example is utterly typical of the amount—and style—of phenotypic change within lineages in the fossil record.

But, though Campbell accurately conveys the sequence of organismic phenotypic change in these trilobites, his diagram utterly distorts the actual pattern of change through time. Campbell's diagram portrays the genera *Acernaspis, Ananaspis, Paciphacops,* and *Phacops.*[10] My reading of the data, while supporting Campbell's characterization of that sequence of ornamental changes, sees a number of species, distributed in various parts of the world, that last through all or significant intervals of conventionally designated divisions of Silurian and Devonian time.[11] I have summarized the data in fig. 7.4.

The point of the pattern that emerges is simple enough. In each stage, or sometimes over an interval of several stages, there is one genus of phacopid trilobite present, with a number of species scattered over the world in appropriate ecosystems. In the Lower Silurian in particular, there are many species of *Acernaspis*, including several subgenera. All share the

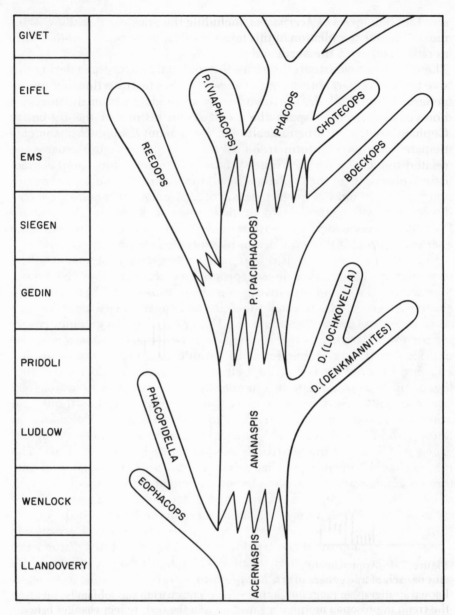

Figure 7.3   Campbell's phylogenetic scheme for Phacopinae (Campbell 1977, fig. 4).

same basic features of *Acernaspis*, including the granular exoskeletal ornament. The minor radiation of this taxon involves a variety of features (such
as conformation of the head, and size and shape of the schizochroal eyes).
The pattern is consistent: the radiation of each genus within each successive time increment does not involve the various features that vary *among*
genera (i.e., the features involved in the "trends") within the sixty-mil-
lion-year period that spans the bulk of the history of the subfamily.
Cephalic anatomy can be radically rearranged, but the features that "participate" in the long-term trend remain constant among those closely
related species within each time increment. It is as if there is a "standard
issue" phacopid trilobite—suitable for occupation of many of the world's
shallow-water marine communities—in each division of geological time

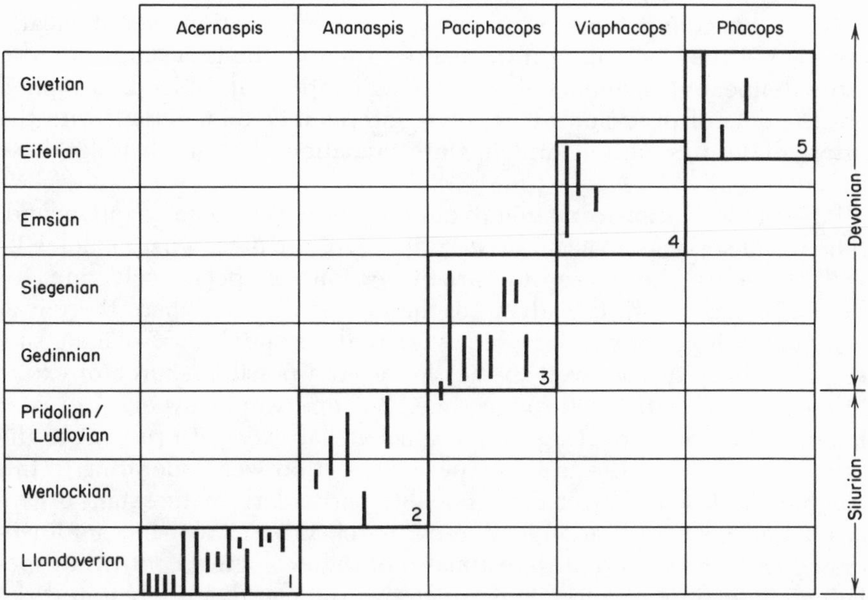

Figure 7.4 Approximate stratigraphic ranges of some of the better-known species of each of five genera of trilobites of the Subfamily Phacopinae. Boxes encase
known stratigraphic ranges of each genus. Trends within the subfamily, including
the trend in prosopon morphology discussed in the text, reflect changes *between*,
rather than *within*, the five boxes. In the case of the trend in prosopon (ornament),
the major changes were between boxes 1–2 and 3–4. (1) Species of *Acernaspis*
from North America only. Other species are known from other continents. (2) Species of *Ananaspis* from North America and Europe. (3) Species of *Paciphacops* from
Bolivia (straddling stratigraphic boundary), Australia, and North America. Others
are known from eastern Soviet Union. (4) Species of *Viaphacops* from Bolivia and
North America only. (5) Three of many species of *Phacops* from Europe and North
America only. Ranges occupying midsection of a stage are less precise than other
ranges.

from the Lower Silurian on up through the Middle Devonian. The genera, while apparently largely monophyletic, are for the most part defined and recognized as Campbell's diagram implies, that is, as "stages of evolution" of the ornamental and coordinate anatomical trends. Temporal durations of each species vary, of course. *Paciphacops logani*, from the Lower Devonian of eastern North America, spans the interval Gedinnian–Siegenian, some fourteen million years (Eldredge 1973). Temporal durations of other species seem to be on that order or, in some instances, rather less.

With one exception known to me (and with the interdigitating marks on Campbell's diagram notwithstanding) there is no stratigraphic overlap between species with different states of the ornamental trend.[12] The one exception is an apparently uppermost-Silurian species (as yet undescribed) of *Paciphacops* from Bolivia. Thus, we have an almost absurdly slow rate of change, a rather minor yet persistent trend that consumes some sixty million years of time. Campbell's diagram (fig. 7.3)accurately portrays the conventional synthesis-derived explanation, yet it clearly glosses over the particulars of the matter, treating the taxa simply as arbitrarily delineated segments of an evolving continuum. Indeed, Campbell robs the data of precision by implying interconnectedness (i.e., with the pattern of interdigitation, implying intergradation) that simply is not there in the data.

If the phyletic transformation of entire lineages on some adaptive landscape won't serve as an accurate description of the data, what model will? Well, there are the numerous variant versions of species selection, for one. Particularly with the advent of the Bolivian taxon, where there is at least one example of stratigraphic overlap, differential more-making success of differently endowed species looms as a possible source of explanation of the net trend through time. But one would expect a greater amount of temporal overlap among species that vary with respect to the features involved in the trend if species selection were operating in this example. And, with better data (possible particularly with extant organisms in lineages with a good fossil record—cf. Vrba 1980—but not forthcoming with these trilobites), estimates of the ecological controls of speciation and extinction rates could be drawn sufficiently closely to test the application of Vrba's effect hypothesis on these data.

But there is more to the data than meets the eye—as I have presented the case so far. Both hierarchy and synthesis freely admit that lineages do not evolve in a vacuum, yet we persist in analyzing the phylogenetic histories of monocultures, be they clonal descendants of a single bacterium or the evolutionary histories of entire phyla, as if they do evolve in vacuo. They don't, as this particular case of an evolutionary trend within the Phacopinae amply demonstrates.

We must ask what forms the subdivisions of the geological time scale— the boundaries for the periods and epochs, the systems, series, and stages? The answer is events in the history of life, specifically episodes of extinction and proliferation. The divisions were documented empirically, for the most part by creationist-minded geologists working in a pre-Darwinian intellectual world. And when we examine the rest of the biota, the components of marine communities other than phacopine trilobites, some simple additional facts emerge.

When each of the genera of phacopines in the lineage disappeared from the record, it did so in concert with many taxa of brachiopods, snails, clams, echinoderms, and other trilobites. These nearly simultaneous disappearances are the events—cross-genealogical extinctions—that form the divisions of geological time in the first place. Two further items are noteworthy: in some other trilobite lineages (not to mention brachiopods and other invertebrates), there are also among-generic trends that conform exactly to the phacopine pattern. Genera radiate within the same time interval, but the characters involved in the coeval interspecific variation are not the same as those that vary among genera through time— that is, those characteristics that define the trend. Yet the taxa within these trending lineages appear and disappear (at roughly the same time), at the ends and beginnings of standard geological divisions, respectively. Homalonotid trilobites of the *Trimerus–Dipleura* lineage (spanning the Upper Silurian–Middle Devonian interval) are a case in point. The celebrated trends in the Silurian articulate brachiopod *Eocoelia* (e.g., Ziegler 1966) may well constitute another.

But then again, not all lineages appear to show such trends. Some genera make it through (i.e., the taxa exhibit no conspicuous change across the temporal boundaries), while the net pattern of change through time within still other lineages seems not so much directional as oscillatory or haphazard. There seem to be no particular net directional changes in morphology, for example, in Siluro–Devonian calymenid trilobites, while Dalmanitinae seem repeatedly to explore a variety of anatomical themes, without any one dominant, directional theme throughout the same temporal interval. Thus, only some of the many synchronic lineages living in shallow-water epicontinental marine environments in the Middle Paleozoic exhibit directional morphological trends. Yet generic turnover among all these lineages is exceptionally high at these subdivisional boundaries within the main geological systems.

What can such data imply? The best description of phacopine trends, I believe, sees the events in ecological context. Many phacopine species at any one time had representative populations in a variety of ecological settings at any one point in time. There was speciation and extinction throughout the interval, and in some instances (particularly the Lower Silurian) there was considerable phenotypic diversification. Judging by events in other lineages, there were occasional episodes of cross-genealogical extinction to a degree severe enough to affect nearly all species of particular lineages worldwide—insofar as the data show. When the ecosystems were reestablished, some lineages came back in modified form, repopulated from a substock that was manifestly not dominant (but did exist, as the single Bolivian species of *Paciphacops* seems to show for phacopines). Many other lineages came back unchanged, or at least unchanged with respect to any putative long-term trends. The null hypothesis—one that remains untested (and not at all considered under the synthesis) yet is very definitely empirically vulnerable—is that in the reestablishment of ecological biotal systems following global extinction events, some lineages come back further modified in the same direction of a prevailing trend already accumulating within the lineage; the majority come back unmodified or modified in some manner not perceptibly involved in some linear trend. There should be a regular distribution of such anatomical patterns, arguably a normal distribution, similar at all cross-genealogical extinction events of comparable magnitude. We tend, after all, to focus on the trends, the anatomical modifications, and to ignore the larger percentage of examples that show little, or fluctuating, change.

Such a description says that trends—even those that are carried on over some sixty million years—are merely examples culled from a spectrum of possibilities. If it is so that ecosystems involving a variety of lineages remain stable (with just the usual background extinction and speciation ticking away) until some relatively major cross-genealogical, ecological event intensifies extinctions, then it seems reasonable to suppose on chance alone that some of the lineages that reappear in modified form will be modified in the same direction as that of a trend already established in the lineage. Others remain unmodified, and others change in some other directions vis-à-vis earlier changes within the lineage. One very nice aspect of the hypothesis is that we can go to the fossil record and see if such a spectrum actually exists. Nobody has done so yet, because the question hasn't been asked. We have been far too busy looking for explanations of trends couched in the expressive language of progressive adaptation to bother framing the question this way. Readers of Simpson's (1944, pp. 90–93, 209–10, the former repeated virtually verbatim in Simpson 1953, pp. 157–59) discussions of equid evolution in terms of adaptive landscapes and adaptive zones will recognize the usual mode of explanation of evolutionary trends prevalent in the synthesis. Yet the data of equid evolution almost assuredly will look very similar to those presented here for trends in phacopine trilobite evolution—when those data are finally compiled.

A final word on trends: none of the above is to be construed somehow as anti-adaptation. The organisms in each of those extinct species were more or less fit because they all more or less possessed phenotypes adapted to a particular mode of life. That is the case for organisms today, and we simply assume that it was always thus. And natural selection, one can only assume, was the force that shaped the modifications of those adaptations through time. But the fossil record supplies an additional piece of insight: all adaptations are good enough for the moment, and moments can stretch out to millions of years. It takes the higher-level degradation of large-scale ecological units to reset the evolutionary clock. With no mass extinctions, marine or terrestrial, life would still look pretty much as it did in the Devonian.

## SOME FINAL THOUGHTS

When the masters of the synthesis—Dobzhansky, Mayr, and Simpson—were writing, they were well aware of the notion that evolution, meaning change, is somehow inevitable. Dobzhansky saw the idea and tried to avoid it, while at least in one passage Simpson was openly enthusiastic about the inevitability of change—and for good reason. Not only has there been a long and deep intellectual history embracing precisely this point, but the very act of framing a believable theory of how evolutionary change can come about, if the theory is any good at all, ought to turn the tables from "evolution is difficult to accomplish" to "I don't see how evolutionary change can be avoided." Darwin himself stressed this very spirit.

Yet, looking at the history of life from the top down, so to speak (meaning from the relatively coarse-grained perspective of a paleontologist), evolution seems to me much more a matter of producing workable systems—organisms that (1) can function in the economic sphere and (2) can reproduce. Once the system is up and running, it will do so indefinitely—until something happens. Nearly always, that something is physicochemical environmental change. The economic game is disrupted. Most often, as the fossil record so eloquently tells us, the system is downgraded and must be rebuilt, using the survivors to fashion the workable new version. At other times, new economic situations are simply opened up, as in the rise of $O_2$ tension (through marine photosynthesis). And, yes, occasionally better mousetraps do seem to be built, though the history of adaptation is much more commonly the other way around: the mousetrap is invented that allows a new way of succeeding in the biological economy, and the tens of millions of years of subsequent variation are but themes and variations—a notion developed, for example, by Simpson (1959a) as "key innovations."

We are accustomed to thinking of evolution as change, but it is only partly that. It is, at base, a matter of making a living and the transmission of the instructions for doing so to the next generation. The conservative retention of both systems is the prime order of the day, day in and day out, for millions and, ultimately, billions of years. But if we insist upon defining evolution strictly as change and further focus our notions of change on adaptive modification, we will have done worse than merely distort the nature of life's history. Such a picture of the adaptive history of life forces us to see species as adaptive devices, as classlike genealogical entities that are meaningfully understood to have direct roles to play in the economy of life. If we revamp our ontology, as I believe the exigencies of reality simply force us to do, we see that neither view is "true." The synthesis, to the extent that it did embrace phenomena beyond the within-species level, insisted on such a view of species. And that is its main fault, its paramount failing as an evolutionary theory.

One very straightforward consequence of adopting some version of a hierarchical approach to evolution is that a far broader set of phenomena can be acknowledged as directly relevant to evolutionary biology than has been customarily the case. The kinetics of evolution are no longer seen to be confined to the processes we study in a *Drosophila* population cage, though it is even more important to reiterate that such classes of events in selection experiments are no less relevant now than they ever were. There has been this persistent belief that to acknowledge that evolution is more than a matter of deterministic and random processes of allelic shuffling somehow implies that such processes are thereby rendered superfluous.

Hierarchy allows us to acknowledge that both ecological and genealogical entities, events, and processes are involved in the "evolutionary process." The entities all appear to be stable individuals. They are hierarchically arranged. There are processes intrinsic to each level that are not reducible to those of lower levels—or subsumed by higher levels. The genealogical processes are birth and death; differential births and deaths go on at more than the conventional among-organism, within-deme level. There may be true interdemal and interspecific selection, as well as biased replication of genomic elements independent of natural selection and genetic drift (cf. Dover 1982). In any case, descriptive patterns of differential births and deaths of higher-level individuals—notably species—surely exist and form a major feature of the history of life. The geometry of those past patterns, cumulatively, has everything to do with the present configuration of life, from both an ecological and a genealogical point of view. Whether the biases in births and deaths of species arise from processes intrinsic to that level (species selection) or as a reaction (as I believe) to higher-level ecological phenomena that are translated to the genealogical system as a series of effects (up from the organismic level) remains to be seen. Epistemologically, hierarchy theory is a formal embodiment of the principle that evolution is probably a more complex affair than the synthesis would have us believe. Hierarchy offers us, I think, a more realistic, hence more useful, framework for the investigation of this complexity.

The bare bones of such a hierarchical theory along the lines that I have sketched is indeed complex, but manageably so. Nothing—literally no thing, no entity—exists in isolation from other entities in either of the two hierarchical process systems. It is possible to describe one aspect of the dynamic—say, the effect of the environment on inducing adaptive change—and forget that it is the nested system of genetic information that determines in the first place the actual biotic nature of that environment.

Perhaps the greatest snare and delusion of such a hierarchical approach is that we come dangerously close to saying that all things are true. But evolutionary theory *has* been like the blind man and the elephant: each of the various disciplines has been looking at its own part of the elephant and claiming that the system in general resembles its particular piece. All things are *not* true, but all the parts of the elephant are surely relevant to understanding what an elephant is. If that is the main advantage to be gained from looking at life as a series of hierarchical systems and evolution as, consequently, a multilevel, richly complex process, it is enough.

## NOTES

1. Indeed, the confusion between the ecological and genealogical hierarchies is very familiar to me, as my initial interest in hierarchies stems from arguments with S. N. Salthe, whose paper (Salthe 1975) described a hierarchy of levels (molecular, cellular–organismic, population, community). I thought that the notion of hierarchy was valuable, but the hierarchy I associated with evolution involved genes, organisms, species, and monophyletic taxa. Persistent joint failure to integrate these two systems into a single, coherent framework led to our description (Eldredge and Salthe 1984) of two parallel hierarchies, a modified version of which I develop here. But confusion runs deep: the hierarchy I described (Eldredge 1982b) includes regional biotas above monophyletic taxa (a vain attempt to incorporate cross-genealogical phenomena such as extinctions) and amounts to mixing apples and oranges.

2. The one possible exception to the generalization that a species cannot be an ecological interactor is when it is coextensive with a single deme and the deme itself nearly constitutes a single population.

3. Populations and demes also die through the deaths of component organisms, but here the reduction is more trivial. It *is* realistic in a sense to speak of the deaths of populations and communities directly, as, for example, when low tides of abnormally long duration kill off such entities in the rocky intertidal zone or when, more permanently, the sea level changes and the locus of such zones simply moves. Here the abiotic realm simply removes the possibility for such entities to continue to exist—a direct effect at the higher level, though death is still achieved by the removal of each and every organism within each population or community.

4. Note that, should such a hypothetical macromutation be induced by a cosmic ray, the environment is still further implicated, though the connection remains remote and indirect, the hypothetical phenotypic effect not in any case being an adaptation to cosmic rays.

5. No taxon, including kingdoms, needs to include more than a single species, nor does a species need to include more than one deme. In such circumstances, of course, the genealogical hierarchy stops at the deme level.

6. I have discussed this matter of habitat shift and its tracking (habitat recognition) at length elsewhere (Eldredge 1985). Here I only note that habitat recognition probably amounts to the same thing as stabilizing or centripetal selection. Organisms that fall within the range of "normalcy" in terms of possessing the requisite phenotypic properties for appropriate habitat recognition and niche exploitation (plus conspecific mate recognition) are just that—normal. Though technically correct, calling the pruning process of organisms that fail to fit these requirements "selection" nonetheless does highlight the "selection does all" attitude that the synthesis came to adopt so thoroughly. "Habitat recognition" carries with it more of a flavor of neutrality than of definitive selection.

7. This assumes a lot, of course, but the origin of life historically is beyond the purview of standard evolutionary theory. This is for good reason: stasis and change in biological systems are inherently—and fundamentally—different problems from the development of biotic from abiotic systems.

8. At a conference on hierarchy at Yale University (April 1983), an exchange between S. N. Salthe, whose viewpoint is staunchly hierarchical (e.g., Salthe 1975, 1985; Eldredge and Salthe 1984), and some non-hierarchically inclined participants illustrated precisely this point. The issue was the adaptive significance of the penis (mammalian, presumably). There is no question that it is an adaptation. The question is, who or what benefits from its use? Conventional perspective sees the issue as trivial: the penis is ideally suited for the perpetuation of genes. The genes benefit, so there is no problem in seeing the penis as an adaptation. On the other hand, Salthe saw the penis as a means of perpetuation of what he called the population; the population can be construed to benefit equally as well as the genes. (No one mentioned organisms.) Salthe's hierarchical perspective allowed him to add to the argument; neither position is more—or less—correct than the other.

9. For the record, the bipartite 3p glabellar furrows, with the distal tip of the distal moiety not in communication with the axial furrow, are an absolutely unique feature among trilobites. This synapomorphy links all Lower Silurian–Upper Devonian Phacopinae.

10. Campbell sees *Viaphacops* as an offshoot of the main lineage, though members of this genus are indeed anatomically intermediate between *Paciphacops* and *Phacops* in some respects, including some aspects of the ornamental features forming the trend that I am discussing here. *Eophacops*, however, lacks the very synapomorphies (e.g., bipartite 3p furrow not communicating with the axial furrow) that unite the subfamily, hence does not even belong on Campbell's diagram. More to the point, the very style of depicting one genus as ancestral to another bespeaks grades, means that the genera are not monophyletic, and requires a theory for the origin of genera not available even in the synthesis. The irony here is that these sequential genera probably really *are* monophyletic, or nearly so. They are merely not linked up as a single coherent lineage, with each genus "becoming extinct" by evolving into the next advanced form, as Campbell shows in his diagram. One final point: some of the branching portrayed in Campbell's diagram involves taxa *not* involved in the trend under discussion here. Specifically, these taxa retain the primitive (plesiomorphic) traits. For example, *Reedops* retains an ornament of finely disseminated granules—and lived during Early Devonian times.

11. Traditional in geological usage, the Silurian and Devonian are periods of time divided into epochs (e.g., Early, Middle, and Late). However, all the rocks formed during these periods are considered systems, and systems are divided into series (e.g., Lower, Middle and Upper). Divisions of series are stages. For the Silurian, the standard stages are (from the lowest) Llandoverian, Wenlockian, Ludlovian, and Pridolian. For the Devonian, the sequence begins with the Gedinnian, then runs through Siegenian, Emsian, Eifelian, and Givetian (Middle Devonian)—the sequence of stages that embraces the trend I am discussing.

12. I agree with Campbell (1977) that younger species, at least, of *Viaphacops* (which he sees as a subgenus of *Paciphacops*) have autapomorphies removing them from consideration as direct ancestors of *Phacops*. Thus, the overlap in stratigraphic range between these taxa (boxes 4 and 5 of fig. 7.4) is of less significance in the analysis of trends than the occurrence of the intermediate Bolivian species that effectively links *Ananaspis* with *Paciphacops*.

# References

Alexander, R. D., and G. Borgia. 1978. Group selection, altruism, and the levels of organization of life. *Ann. Rev. Ecol. Syst.* 9:449–74.

Allen, T. H. F., and T. B. Starr. 1982. *Hierarchy: Perspectives for Ecological Complexity.* Chicago: University of Chicago Press.

Altokhov, Y. P. 1982. Biochemical population genetics and speciation. *Evolution* 36:1168–81.

Alvarez, L. W., W. Alvarez, F. Asaro, and H. V. Michel. 1980. Extraterrestrial cause for the Cretaceous-Tertiary extinction. *Science* 208:1095–1108.

Arnold, A. J., and K. Fristrup. 1982. The theory of evolution by natural selection: a hierarchical expansion. *Paleobiology.* 8:113–29.

Ayala, F. 1975. Genetic differentiation during the speciation process. *Evol. Biol.* 8:1–78.

Bernstein, H., H. C. Byerly, F. A. Hopf, and R. E. Michod. In press. Sex and the emergence of species. *J. Theoretical Biology.*

Bock, W. J. 1959. Preadaptation and multiple evolutionary pathways. *Evolution* 13:194–211.

————. 1979. The synthetic explanation of macroevolutionary change—a reductionistic approach. In J. H. Schwartz and H. B. Rollins, eds., *Models and Methodologies in Evolutionary Theory, Bull. Carnegie Mus. Nat. Hist.* 13:20–69.

Boucot, A. J. 1978. Community evolution and rates of cladogenesis. *Evol. Biol.* 11:545–655.

Bradshaw, J. S. 1961. Laboratory experiments on the ecology of Foraminifera. *Cushman Found. Foram. Res. Contr.* 12:87–106.

Bretsky, P. W. 1968. Evolution of Paleozoic marine invertebrate communities. *Science* 159:1231–33.

————. 1969. Evolution of Paleozoic benthic marine invertebrate communities. *Palaeogr. Palaeoclimatol. Palaeoecol.* 6:45–59.

Bretsky, P. W., and D. M. Lorenz. 1969. Adaptive response to environmental stability: a unifying concept in paleoecology. *Proc. North Amer. Paleont. Conv.,* part E: 522–50.

Bunge, M. 1977. *Treatise of Basic Philosophy. Vol. 3: The Furniture of the World.* Dordrecht, Netherlands: D. Reidel Publishing Company.

Bush, G. L. 1975. Modes of animal speciation. *Ann. Rev. Ecol. Syst.* 6:339–64.

Campbell, K. S. W. 1967. Henryhouse trilobites. *Bull. Okla. Geol. Surv.* 115:1–68.

————. 1977. Trilobites of the Haragan, Bois d'Arc, and Frisco Formations (Early Devonian) Arbuckle Mountains region, Oklahoma. *Bull. Okla. Geol. Surv.* 123:1–227.

Campbell, T. 1974. "Downward causation" in hierarchically organised biological systems. In F. J. Ayala and Th. Dobzhansky, eds., *Studies in the Philosophy of Biology*, 179–86. San Franscisco: University of California Press.

Carson, H. L. 1981. Untitled letter on Chicago macroevolution conference. *Science* 211:773.

———. 1982. Speciation as a major reorganization of polygenic balances. In *Mechanisms of Speciation*, 411–33. New York: Alan R. Liss.

Charlesworth, B., R. Lande, and M. Slatkin. 1982. A neo-Darwinian commentary on macroevolution. *Evolution* 36:474–98.

Cohen, J. A., and H. J. Brockmann. 1983. Breeding activity and mate selection in the horseshoe crab, *Limulus polyphemus*. *Bull. Marine Science* 33:274–81.

Coon, C. S. 1962. *The Origin of Races*. New York: Knopf.

Cracraft, J. 1982. A nonequilibrium theory for the rate-control of speciation and extinction and the origin of macroevolutionary patterns. *Syst. Zool.* 31:348–65.

Damuth, J. In press. "Species" selection: a reformulation in terms of natural functional units. *Evolution*.

Darwin, C. 1859. *On the Origin of Species*. Facsimile ed. New York: Atheneum, 1967.

Dawkins, R. 1976. *The Selfish Gene*. Oxford: Oxford University Press.

———. 1982. *The Extended Phenotype: The Gene as the Unit of Selection*. Oxford and San Francisco: W. H. Freeman.

Dobzhansky, T. 1935. A critique of the species concept in biology. *Philosophy of Science* 2:344–55.

———. 1937a. *Genetics and the Origin of Species*. Reprint ed. New York: Columbia University Press, 1982.

———. 1937b. Genetic nature of species differences. *Amer. Naturalist* 71:404–20.

———. 1940. Speciation as a stage in evolutionary divergence. *Amer. Naturalist* 74:312–21.

———. 1941. *Genetics and the Origin of Species*. Second ed. New York: Columbia University Press.

———. 1951. *Genetics and the Origin of Species*. Third ed. New York: Columbia University Press.

———. 1970. *Genetics of the Evolutionary Process*. New York: Columbia University Press.

———. 1973. Nothing in biology makes sense except in the light of evolution. *Amer. Biol. Teacher* 35:125–29.

Dobzhansky, T., F. J. Ayala, G. L. Stebbins, and J. W. Valentine. 1977. *Evolution*. San Francisco: W. H. Freeman.

Doolittle, W. F., and C. Sapienza. 1980. Selfish genes, the phenotype paradigm and genome evolution. *Nature* 284:601–3.

Dover, G. A. 1982. Molecular drive: a cohesive mode of species evolution. *Nature* 299:111–17.

Eldredge, N. 1971. The allopatric model and phylogeny in Paleozoic invertebrates. *Evolution* 25:156–67.

———. 1973. Systematics of Lower and Lower Middle Devonian species of the trilobite *Phacops* Emmrich in North America. *Bull. Amer. Mus. Nat. Hist.* 151:285–338.

———. 1979. Alternative approaches to evolutionary theory. In J. H. Schwartz and H. B. Rollins, eds., *Models and Methodologies in Evolutionary Theory*, *Bull. Carnegie Mus. Nat. Hist.* 13:7–19.

————. 1982a. Introduction (for reprinted edition of Mayr 1942), xv-xxxvii. New York: Columbia University Press.

————. 1982b. Phenomenological levels and evolutionary rates. *Syst. Zool.* 31:338–47.

————. 1982c. *The Monkey Business: A Scientist Looks at Creationism.* New York: Washington Square Press (Pocket).

————. 1984. Simpson's inverse: bradytely and the phenomenon of living fossils. In N. Eldredge and S. M. Stanley, eds., *Living Fossils*, 272–77. New York: Springer Verlag.

————. 1985. *Time Frames.* New York: Simon and Schuster.

Eldredge, N., and J. Cracraft. 1980. *Phylogenetc Patterns and the Evolutionary Process: Method and Theory in Comparative Biology.* New York: Columbia University Press.

Eldredge, N., and S. J. Gould. 1972. Punctuated equilibria: an alternative to phyletic gradualism. In T. J. M. Schopf, ed., *Models in Paleobiology*, 82–115. San Francisco: Freeman, Cooper.

————. 1974. Reply to Hecht. *Evol. Biol.* 7:303–8.

Eldredge, N., and S. N. Salthe. 1984. Hierarchy and evolution. In R. Dawkins and M. Ridley, eds., *Oxford Surveys in Evol. Biology* 1:182–206.

Eldredge, N., and S. M. Stanley, eds. 1984. *Living Fossils.* New York: Springer Verlag.

Eldredge, N., and I. Tattersall. 1982. *The Myths of Human Evolution.* New York: Columbia University Press.

Fisher, R. A. 1930. *The Genetical Theory of Natural Selection.* Oxford: Clarendon Press.

Fowler, C. W., and J. A. MacMahon. 1982. Selective extinction and speciation: their influence on the structure and functioning of communities and ecosystems. *Amer. Naturalist* 119:480–98.

Fox, L. R., and P. A. Morrow. 1981. Specialization: species property or local phenomenon? *Science* 211:887–93.

Futuyma, D. J. 1979. *Evolutionary Biology.* Sunderland, Mass.: Sinauer.

Futuyma, D. J., and M. Slatkin, eds. 1983. *Coevolution.* Sunderland, Mass.: Sinauer.

Ghiselin, M. T. 1966. On psychologism in the logic of taxonomic controversies. *Syst. Zool.* 15:207–15.

————. 1974a. A radical solution to the species problem. *Syst. Zool.* 23:536–44.

————. 1974b. *The Economy of Nature and the Evolution of Sex.* Berkeley and Los Angeles: University of California Press.

Goldschmidt, R. 1940. *The Material Basis of Evolution.* Reprint ed. New Haven: Yale University Press, 1982.

Goodwin, B. C. 1982. Development and evolution. *J. Theoretical Biol.* 97:43–55.

Gorczynski, R. M., and E. J. Steele. 1981. Simultaneous yet independent inheritance of somatically acquired tolerance to two distinct H-2 antigenic haplotype determinants in mice. *Nature* 289:678–81.

Gould, S. J. 1977. *Ontogeny and Phylogeny.* Cambridge: Belknap (Harvard University Press).

————. 1980a. Is a new and general theory of evolution emerging? *Paleobiology* 6:119–30.

————. 1980b. G. G. Simpson, paleontology, and the modern synthesis. In E. Mayr and W. B. Provine, eds., *The Evolutionary Synthesis: Perspectives on the Unification of Biology*, 153–72. Cambridge: Harvard University Press.

————. 1982a. Introduction (to reprint edition of Dobzhansky 1937a), xvii-xli. New York: Columbia University Press.

————. 1982b. Darwinism and the expansion of evolutionary theory. *Science* 216:380–87.

Gould, S. J., and N. Eldredge. 1977. Punctuated equilibria: the tempo and mode of evolution reconsidered. *Paleobiology* 3:115–51.

Gould, S. J., and E. S. Vrba. 1982. Exaptation—a missing term in the science of form. *Paleobiology* 8:4–15.

Greenacre, M. J., and E. S. Vrba. 1984. A correspondence analysis of biological census data. *Ecology* 65: 984–97.

Grene, M. 1959. Two evolutionary theories. *Brit. J. Phil. Sci.* 9:110–27.

Haldane, J. B. S. 1932. *The Causes of Evolution.* New York: Harper.

Hamilton, W. D. 1964a. The genetical theory of social behavior. I. *J. Theoretical Biol.* 7:1–16.

Hamilton, W. D. 1964b. The genetical theory of social behavior. II. *J. Theoretical Biol.* 7:17–32.

Hapgood, F. 1984. The importance of being Ernst. *Science 84* 5:40–46.

Haugh, B. N., and B. M. Bell. 1980. Fossilized viscera in primitive echinoderms. *Science* 209:653–57.

Hecht, M. K., and B. Schaeffer, eds. 1965. Symposium: the origin of higher levels of organization. *Syst. Zool.* 14:245–342.

Hennig, W. 1966. *Phylogenetic Systematics.* Urbana: University of Illinois Press.

Ho, M. W., and P. T. Saunders 1979. Beyond neo-Darwinism: an epigenetic approach to evolution. *J. Theoretical Biol.* 78:673–91.

Hull, D. L. 1974. *Philosophy of Biological Science.* Englewood Cliffs, N.J.: Prentice-Hall.

————. 1976. Are species really individuals? *Syst. Zool.* 25:174–91.

————. 1978. A matter of individuality. *Phil. Sci.* 45:335–60.

————. 1980. Individuality and selection. *Ann. Rev. Ecol. Syst.* 11:311–32.

Huxley, J. S. 1941. Evolutionary genetics. *Proc. Eighth Int. Genet. Congr., Edinburgh* 1939:157–64.

———— 1942. *Evolution: The Modern Synthesis.* New York: Harper.

———— 1958. Evolutionary processes and taxonomy with special reference to grades. *Uppsala Univ. Arssk.* 1958:21–38.

Johnson, R. G. 1972. Conceptual models of benthic marine communities. In T. J. M. Schopf, ed., *Models in Paleobiology*, 148–159. San Francisco: Freeman, Cooper.

Kimura, M. 1968. Evolutionary rates at the molecular level. *Nature* 217: 624–26.

Kottler, M. J. 1983. A history of biology: diversity, evolution, inheritance [review of Mayr, 1982]. *Evolution* 37:868–72.

Levene, H., L. Ehrman, and R. Richmond. 1970. Theodosius Dobzhansky up to now. In M. K. Hecht and W. C. Steere, eds., *Essays in Evolution and Genetics in Honor of Theodosius Dobzhansky* (supplement to *Evol. Biol.*), 1–41. New York: Appleton-Century-Crofts.

Levins, R. 1968. *Evolution in Changing Environments*. Princeton: Princeton University Press.

Levinton, J. S. 1983. Stasis in progress: the empirical basis of macroevolution. *Ann. Rev. Ecol. Syst.* 14:103–37.

Levinton, J. S., and C. M. Simon. 1980. A critique of the punctuated equilibria model and implications for the detection of speciation in the fossil record. *Systematic Zoology* 29:130–42.

Lewin, R. 1980. Evolutionary theory under fire (report on 1980 Chicago conference on macroevolution). *Science* 210:883–87.

Lewontin, R. C. 1962. Interdeme selection controlling a polymorphism in the house mouse. *Amer. Naturalist* 96:65–78.

————. 1974. *The Genetic Basis of Evolutionary Change*. New York: Columbia University Press.

————. 1978. Adaptation. *Sci. American* 239:212–30.

————. 1980. Theoretical population genetics in the evolutionary synthesis. In E. Mayr and W. B. Provine, eds., *The Evolutionary Synthesis*, 58–68. Cambridge: Harvard University Press.

Lewontin, R. C., and L. C. Dunn. 1960. The evolutionary dynamics of a polymorphism in the house mouse. *Genetics* 45:705–22.

Lewontin, R. C., and J. L. Hubby. 1966. A molecular approach to the study of genic heterozygosity in natural populations. II. Amount of variation and degree of heterozygosity in natural populations of *Drosophila pseudoobscura*. *Genetics* 54:595–609.

Lewontin, R. C., J. A. Moore, W. B. Provine, and B. Wallace, eds. 1981. *Dobzhansky's Genetics of Natural Populations*, I-XLIII. New York: Columbia University Press.

MacMahon, J. A., D. L. Phillips, J. V. Robinson, and D. J. Schimpf. 1978. Levels of biological organization: an organism-centered approach. *Bioscience* 28:700–704.

MacMahon, J. A., D. J. Schimpf, D. C. Andersen, K. G. Smith, and R. L. Bayn Jr. 1981. An organism-centered approach to some community and ecosystem concepts. *J. Theoretical Biol.* 88:287–307.

Margulis, L. 1974. Five-kingdom classification and the origin and evolution of cells. *Evol. Biol.* 7:45–78.

Mather, K. 1941. Variation and selection of polygenic characters. *J. Genetics* 41:159–93.

Maynard Smith, J. 1978. *The Evolution of Sex*. Cambridge: Cambridge University Press.

Mayr, E. 1942. *Systematics and the Origin of Species*. Reprint ed. New York: Columbia University Press, 1982.

————. 1947. Ecological factors in speciation. *Evolution* 1:263–88.

————. 1954. Change of genetic environment and evolution. In J. Huxley, A. C. Hardy, and E. B. Ford eds., *Evolution as a Process*, 157–80. London: Allen and Unwin.

————. 1963. *Animal Species and Evolution*. Cambridge: Belknap (Harvard University Press).

————. 1964. Preface to reprint edition of Mayr, 1942, ix-xi. New York: Dover.

————. 1969. *Principles of Systematic Zoology*. New York: McGraw-Hill..

————. 1970. *Populations, Species, and Evolution*. Cambridge: Harvard University Press.

—————. 1980a. Prologue: some thoughts on the history of the evolutionary synthesis. In E. Mayr and W. B. Provine, eds., *The Evolutionary Synthesis*, 1–48. Cambridge: Harvard University Press.

—————. 1980b. The role of systematics in the evolutionary synthesis. In E. Mayr and W. B. Provine, eds., *The Evolutionary Synthesis*, 123–36. Cambridge: Harvard University Press.

—————. 1980c. Biographical essays: G. G. Simpson. In E. Mayr and W. B. Provine, eds., *The Evolutionary Synthesis*, 452–63. Cambridge: Harvard University Press.

—————, 1982. *The Growth of Biological Thought: Diversity, Evolution, Inheritance.* Cambridge: Belknap (Harvard University Press).

Mayr, E., E. G. Linsley, and R. L. Usinger. 1953. *Methods and Principles of Systematic Zoology.* New York: McGraw-Hill.

Mayr, E., and W. B. Provine, eds. 1980. *The Evolutionary Synthesis: Perspectives on the Unification of Biology.* Cambridge: Harvard University Press.

Miles, R. S. 1969. Features of placoderm diversification and the evolution of the arthrodire feeding mechanism. *Trans. Roy. Soc. Edinburgh* 68:19–170.

Muller, H. J. 1932. Some genetic aspects of sex. *Amer. Naturalist* 66:118–38.

Nelson, G., and N. Platnick. 1981. *Systematics and Biogeography. Cladistics and Vicariance.* New York: Columbia University Press.

Newell, N. D. 1963. Crises in the history of life. *Sci. American* 208:76–92.

Novacek, M. 1984. Evolutionary stasis in the elephant-shrew, *Rhynchocyon.* In N. Eldredge and S. M. Stanley, eds., *Living Fossils*, 4–22. New York: Springer Verlag.

Odum, E. P. 1971. *Fundamentals of Ecology.* Third ed. Philadelphia: Saunders.

Orgel, L. E., and F. H. C. Crick. 1980. Selfish DNA: the ultimate parasite. *Nature* 284:604–7.

Oster, G., and P. Alberch. 1982. Evolution and bifurcation of developmental programs. *Evolution* 36:444–59.

Paterson, H. E. H. 1978. More evidence against speciation by reinforcement. *S. Afr. J. Sci.* 74:369–71.

—————. 1980. A comment on "mate recognition systems." *Evolution* 34:330–31.

—————. 1981. The continuing search for the unknown and unknowable: a critique of contemporary ideas on speciation. *S. Afr. J. Sci.* 77:113–19.

—————. 1982. Perspectives on speciation by reinforcement. *S. Afr. J. Sci.* 78:53–57.

Platnick, N. I., and H. D. Cameron. 1977. Cladistic methods in textual, linguistic, and phylogenetic analysis. *Syst. Zool.* 26:380–85.

Raup, D. M. 1977. Stochastic models in evolutionary paleontology. In A. Hallam, ed., *Patterns of Evolution, as Illustrated by the Fossil Record*, 59–78. New York: Elsevier.

Raup, D. M., and J. J. Sepkoski. 1982. Mass extinctions in the marine fossil record. *Science* 215:1501–3.

Rensch, B. 1947. *Neuere Probleme der Abstammungslehre.* Stuttgart: F. Enke.

—————. 1954. *Neuere Probleme der Abstammungslehre.* Second ed. Stuttgart: F. Enke.

—————. 1960. *Evolution above the Species Level.* New York: Columbia University Press.

Robson, G. C., and O. W. Richards. 1936. *The Variations of Animals in Nature.* London: Longmans, Green.

Salthe, S. N. 1975. Problems of macroevolution (molecular evolution, phenotype definition, and canalization) as seen from a hierarchical viewpoint. *Amer. Zool.* 15:295–314.

————. 1985. *Evolving Hierarchical Systems: Their Structure and Representation.* New York: Columbia University Press.

Schaeffer, B. 1948. The origin of a mammalian ordinal character. *Evolution* 2:164–75.

————. 1965. The role of experimentation in the origin of higher levels of organization. *Syst. Zool.* 14:318–36.

Schindewolf, O. H. 1936. Paläontologie, Entwicklungslehre und Genetik. Berlin: Borntraeger.

Sepkoski, J. J. 1980. *Charts of the Marine and Continental Fossil Record.* Published by the author.

Shapere, D. 1980. Interpretive issues in the evolutionary synthesis., *In* E. Mayr and W. B. Provine, eds., *The Evolutionary Synthesis*, 387–98. Cambridge: Harvard University Press.

Simon, H. A. 1962. The architecture of complexity. *Proc. Amer. Phil. Soc.* 106:467–82.

Simpson, G. G. 1944. *Tempo and Mode in Evolution.* Reprint ed. New York: Columbia University Press.

————. 1949. *The Meaning of Evolution.* New Haven: Yale University Press.

————. 1953. *The Major Features of Evolution.* New York: Columbia University Press.

————. 1959a. The nature and origin of supraspecific taxa. *Cold Spring Harbor Symp. Quant. Biol.* 24:255–71.

————. 1959b. Mesozoic mammals and the polyphyletic origin of mammals. *Evolution* 13:405–14.

————. 1961. *Principles of Animal Taxonomy.* New York: Columbia University Press.

————. 1963. The meaning of taxonomic statements. In S. L. Washburn, ed., *Classification and Human Evolution*, 1–31. Chicago: Aldine.

Sober, E., and R. C. Lewontin. 1982. Artifact, cause and genic selection. *Philosophy of Science* 49:157–80.

Sokal, R. R., and T. J. Crovello. 1970. The biological species concept: a critical evaluation. *Amer. Nat.* 104:127–53.

Spomer, G. G. 1973. The concept of "interaction" and "operational environment" in environmental analyses. *Ecology* 54:200–204.

Stanley, S. M. 1973. Effects of competition on rates of evolution, with special reference to bivalve mollusks and mammals. *Syst. Zool.* 22:486–506.

————. 1975a. A theory of evolution above the species level. *Proc. Nat. Acad. Sci.* 72:646–50.

————. 1975b. Clades versus clones in evolution: why we have sex. *Science* 190:282–83.

————. 1979. *Macroevolution: Pattern and Process.* San Francisco: W. H. Freeman.

————. 1984. Does bradytely exist? In N. Eldredge and S. M. Stanley, eds., *Living Fossils*, 278–80. New York: Springer Verlag.

Stearns, S. C. 1976. Life-history tactics: a review of the ideas. *Quart. Rev. Biol.* 51:3–47.

Stebbins, G. L. 1950. *Variation and Evolution in Plants.* New York: Columbia University Press.

Stebbins, G. L., and F. J. Ayala. 1981. Is a new evolutionary synthesis necessary? *Science* 213:967–71.

Steele, E. J. 1979. *Somatic Selection and Adaptive Evolution.* Toronto: Williams and Wallace.

Teggart, F. J. 1925. *Theory of History.* New Haven: Yale University Press. Reprinted as *Theory and Processes of History,* Berkeley and Los Angeles: University of California Press, 1977.

Thompson, P. 1983. Tempo and mode in evolution: punctuated equilibrium and the modern synthetic theory. *Phil. Sci.* 50:432–52.

Valentine, J. W. 1968. The evolution of ecological units above the population level. *J. Paleont.* 42:253–67.

———. 1969. Patterns of taxonomic and ecological structure of the shelf benthos during Phanerozoic time. *Palaeontology* 12:684–709.

———. 1973. *Evolutionary Paleoecology of the Marine Biosphere.* Englewood Cliffs, N.J.: Prentice-Hall.

Van Valen, L. 1976. Energy and evolution. *Evol. Theory* 1:179–229.

Vrba, E. S. 1980. Evolution, species and fossils: how does life evolve? *S. Afr. J. Sci.* 76:61–84.

———. 1983. Macroevolutionary trends: new perspectives on the roles of adaptation and incidental effect. *Science* 221:387–89.

———. 1984a. Evolutionary pattern and process in the sister-group Alcelaphini-Aepycerotini (Mammalia: Bovidae). In N. Eldredge and S. M. Stanley, eds., *Living Fossils,* 62–79. New York: Springer Verlag.

———. 1984b. What is species selection? *Syst. Zool.* 33:318–28.

Vrba, E. S., and N. Eldredge. 1984. Individuals, hierarchies and processes: towards a more complete evolutionary theory. *Paleobiology* 10:146–71.

Wade, M. J. 1978. A critical review of the models of group selection. *Quart. Rev. Biol.* 53:101–14.

Whewell, W. 1837. *History of the Inductive Sciences.* London: Parker.

White, M. J. D. 1978. *Modes of Speciation.* San Francisco: W. H. Freeman.

Whittaker, R. H. 1975. *Communities and Ecosystems.* Second ed. New York: Macmillan.

Wiley, E. O. 1978. The evolutionary species concept reconsidered. *Syst. Zool.* 27:17–26.

———. 1981. *Phylogenetics: The Theory and Practice of Phylogenetic Systematics.* New York: John Wiley.

Wiley, E. O., and D. R. Brooks. 1982. Victims of history—a nonequilibrium approach to evolution. *Syst. Zool.* 31:1–24.

Williams, G. C. 1966. *Adaptation and Natural Selection: A Critique of Some Current Evolutionary Thought.* Princeton: Princeton University Press.

———. 1975. *Sex and Evolution.* Princeton: Princeton University Press.

Willis, J. C. 1940. *The Course of Evolution.* Cambridge: Cambridge University Press.

Wilson, E. O. 1975. *Sociobiology: The New Synthesis.* Cambridge: Harvard University Press.

Wimsatt, W. C. 1980. Reductionistic research strategies and their biases in the units of selection controversy. In T. Nickles, ed., *Scientific Discovery: Case Studies,* 213–59. Dordrecht, Netherlands: D. Reidel.

Wright, S. 1931. Evolution in Mendelian populations. *Genetics* 16:97–159.

————. 1932. The roles of mutation, inbreeding, crossbreeding, and selection in evolution. *Proc. Sixth Int. Congr. Genetics* 1:356–66.

————. 1940. The statistical consequences of Mendelian heredity in relation to speciation. In J. Huxley, ed., *The New Systematics*, 161–83. Oxford: Clarendon Press.

————. 1945. *Tempo and Mode in Evolution:* a critical review [review of Simpson, 1944]. *Ecology* 26:415–19.

————. 1956. Modes of selection. *Amer. Naturalist* 90:5–24.

————. 1967. Comments on the preliminary working papers of Eden and Waddington. In P. S. Moorehead and M. M. Kaplan, eds., *Mathematical Challenges to the Neo-Darwinian Interpretation of Evolution*, 117–20. Philadelphia: Wistar Institute Press.

Wynne-Edwards, V. C. 1962. *Animal Dispersion in Relation to Social Behaviour.* Edinburgh: Oliver and Boyd.

Ziegler, A. M. 1966. The Silurian brachiopod *Eocoelia hemisphaerica* (J. de C. Sowerby) and related species. *Paleontology* 9:523–43.

Zotin, A. I. 1972. *Thermodynamic Aspects of Developmental Biology.* Basel, Switzerland: S. Karger.

# Index